T0139486

WHY QUARK RHYMES WITH PORK

∼

And Other Scientific Diversions

A collection of offbeat and entertaining primarily nontechnical essays on physics and those who practice it, from eminent theoretical physicist N. David Mermin. Bringing together for the first time all thirty of his columns published in *Physics Today*'s *Reference Frame* series from 1988 to 2009, with updating commentary, this humorous and unusual volume includes thirteen other essays, many of them previously unpublished.

Mermin's lively and penetrating writing illuminates a broad range of topics, from the implications of bad spelling in a major science journal, to the crises of science libraries and scientific periodicals, the folly of scientific prizes and honors, the agony of getting funding, and how to pronounce "quark." His witty observations and insightful anecdotes gleaned from a lifetime in science will entertain physicists at all levels as well as anyone else interested in science or scientists at the turn of the 21st century.

∼

N. David Mermin is Horace White Professor of Physics Emeritus at Cornell University. He is known throughout the scientific world as co-author of *Solid State Physics* ("Ashcroft and Mermin") and for his columns in *Physics Today*. He was awarded the first Julius Edgar Lilienfeld Prize of the American Physical Society "for outstanding contributions to physics" in 1989, and is a member of the American Philosophical Society, the U.S. National Academy of Sciences, and the American Academy of Arts and Sciences.

WHY QUARK RHYMES WITH PORK

AND OTHER SCIENTIFIC DIVERSIONS

~

N. David Mermin
Cornell University

CAMBRIDGE
UNIVERSITY PRESS

CAMBRIDGE
UNIVERSITY PRESS

University Printing House, Cambridge CB2 8BS, United Kingdom

Cambridge University Press is part of the University of Cambridge.

It furthers the University's mission by disseminating knowledge in the pursuit of education, learning and research at the highest international levels of excellence.

www.cambridge.org
Information on this title: www.cambridge.org/ 9781107024304

© Cambridge University Press 2016

First published 2016

Printed in the United States of America by Sheridan Books, Inc.

A catalogue record for this publication is available from the British Library

Library of Congress Cataloguing in Publication data
Names: Mermin, N. David, author.
Title: Why quark rhymes with pork, and other scientific diversions /
N. David Mermin.
Description: Cambridge, United Kingdom; New York, NY : Cambridge University
Press, 2016. | 2016 | Brings together all his columns published in Physics Today,
1988–2009, with thirteen related essays, many previously unpublished. |
Includes bibliographical references and index.
Identifiers: LCCN 2015027874| ISBN 9781107024304 (hb : alk. paper) |
Subjects: LCSH: Physics–Popular works. | Science–Popular works.
Classification: LCC QC71.M373 2016 | DDC 530–dc23
LC record available at http://lccn.loc.gov/2015027874

ISBN 978-1-107-02430-4 Hardback

for Hannah and Sam and Ivo

CONTENTS

~

Preface **xi**

Part One · *Reference Frame* Columns, *Physics Today* 1988–2009 **1**

1· What's wrong with this Lagrangean, April 1988 **3**
2· What's wrong with this library, August 1988 **9**
3· What's wrong with these prizes, January 1989 **16**
4· What's wrong with this pillow, April 1989 **23**
5· What's wrong with this prose, May 1989 **29**
6· What's wrong with these equations, October 1989 **35**
7· What's wrong with these elements of reality, June 1990 **43**
8· What's wrong with these reviews, August 1990 **50**
9· What's wrong with those epochs, November 1990 **57**
10· Publishing in Computopia, May 1991 **67**
11· What's wrong with those grants, June 1991 **75**
12· What's wrong in Computopia, April 1992 **82**
13· What's wrong with those talks, November 1992 **90**
14· Two lectures on the wave–particle duality, January 1993 **97**
15· A quarrel we can settle, December 1993 **103**
16· What's wrong with this temptation, June 1994 **109**
17· What's wrong with this sustaining myth, March 1996 **117**
18· The golemization of relativity, April 1996 **124**
19· Diary of a Nobel guest, March 1997 **131**
20· What's wrong with this reading, October 1997 **139**
21· How not to create tigers, August 1999 **147**
22· What's wrong with this elegance, March 2000 **154**

23· The contemplation of quantum computation, July 2000   **161**
24· What's wrong with these questions, February 2001   **167**
25· What's wrong with this quantum world, February 2004   **174**
26· Could Feynman have said this? May 2004   **180**
27· My life with Einstein, December 2005   **187**
28· What has quantum mechanics to do with factoring?
    April 2007   **195**
29· Some curious facts about quantum factoring,
    October 2007   **201**
30· What's bad about this habit, May 2009   **208**

Part Two · Shedding Bad Habits   217

31· Fixing the shifty split, *Physics Today*, July 2012   **219**
32· What I think about Now, *Physics Today*, March 2014   **227**
33· Why QBism is not the Copenhagen interpretation, lecture,
    Vienna, June 2014   **232**

Part Three · More from Professor Mozart   249

34· What's wrong with this book, unpublished, 1992   **251**
35· What's wrong with these stanzas, *Physics Today*, July
    2007   **257**

Part Four · More to be Said   267

36· The complete diary of a Nobel guest, unpublished,
    1996   **269**
37· Elegance in physics, unpublished lecture, Minneapolis,
    1999   **283**
38· Questions for 2105, unpublished lecture, Zurich, 2005   **296**

Part Five · Some People I've Known   319

39· My life with Fisher, lecture, Rutgers University, 2001   **321**

40· My life with Kohn, 2003, updated 2013   **331**
41· My Life with Wilson, lecture, Cornell University, 2014   **341**
42· My life with Peierls, unpublished lecture, Santa Barbara, 1997   **346**

Part Six · Summing it Up   359

43· Writing physics, lecture, Cornell University, 1999   **361**

Index   **376**

PREFACE

~

Sometime in the mid-1980s Gloria Lubkin, the editor of *Physics Today*, invited me to contribute to a new column of opinion called *Reference Frame*. Earlier that decade I had published two articles in *Physics Today*.[1,2] The first described my successful effort to make the ridiculous word "boojum" an internationally accepted scientific term. The second gave a very elementary way of thinking about Bell's Theorem and its implications for our understanding of quantum mechanics. These apparently suggested to Gloria that I'd make a good columnist.

I wasn't so sure. Having to produce something clever and entertaining at regular intervals was not my style. On the occasions when I'd managed to do it, it seemed like a small miracle, unlikely ever to happen again. So while I didn't say no, I kept stalling. A couple of years went by.

Then one day I discovered that *Physical Review Letters*, the world's most important physics journal, was doing something quite ridiculous that seemed to have escaped the attention of all the physicists I told about it. The absurd policy and the fact that nobody seemed to have noticed it made a good story. Another miracle. I sent the story (Chapter 1) to Gloria and became a columnist, joining a group of *Reference Frame* writers that included

---

1 "E Pluribus Boojum: The Physicist as Neologist," *Physics Today*, April 1981, 46–53.
2 "Is the Moon There When Nobody Looks? Reality and the Quantum Theory," *Physics Today*, April 1985, 38–47.

Phil Anderson, David Gross, Leo Kadanoff, Dan Kleppner, Jim Langer, and Frank Wilczek.

After that Gloria would phone every few months requesting more miracles. Somehow she managed to induce them. I came to regard her as my Muse. For 21 years she extracted essays I didn't know were in me. She criticized first drafts and negotiated final versions. As some of these essays reveal, my relations with editors have often been tense, but working with Gloria was always a pleasure. She knew exactly how to do her job, and she knew how to get me to do mine.

In 2009 Gloria Lubkin retired from *Physics Today* and the *Reference Frame* columns came to an end. I found to my surprise that I had produced thirty of them—one every eight issues. Not all were miracles, but surprisingly many were. As I traveled around the world of physics after 1988, giving talks at universities and conferences, I discovered that I was becoming better known for my columns than for my technical scientific papers or textbooks. People wanted to talk to me about the columns. And they remembered some from years earlier.

In 2011 I decided to put them all into a book, and Simon Capelin at Cambridge University Press agreed to publish it. My plan was to write a foreword and an afterword for each of the thirty, setting the stage, clarifying the historical context, providing pedagogical background for the occasional technical ones, and describing the letters they elicited to *Physics Today* and to me. This project started off well and then bogged down. Producing thirty such overviews was less fun and was taking longer than I had anticipated. In 2012 everything ground to a halt. I had found an interpretation of quantum mechanics—the QBism of Chris Fuchs and Rüdiger Schack—that finally made sense of the subject. I set the book aside to write about my epiphany.

In 2014 I ran into an old friend, Leonid Levitov, who had been a postdoc at Cornell as my career as a columnist was getting under-way, and is now a professor at MIT. He lamented the disappearance of *Reference Frame* and asked whether I had thought of collecting my columns into a book. I explained that indeed I had, but I found it harder than I expected to write forewords and afterwords for each column. He shook his head. "David," he reprimanded me, "poems do not have forewords and afterwords." Just like that the shackles that had tied me down for three years crumbled into dust. I told Fuchs and Schack that I was taking a sabbatical from quantum foundations, turned down all invitations to write and lecture about QBism, and finished my book.

My paralyzing, unpoetic forewords and afterwords have all condensed into brief postscripts. I have added to the columns in Chapters 1–30 thirteen closely related chapters. Chapters 31–33 describe my book-delaying QBist epiphany. It was foreshadowed by my final *Reference Frame* column, Chapter 30, which would be incomplete without them. Chapters 34 and 35 contain two more major literary efforts of my alter ego, Professor Mozart, which *Physics Today* either declined to publish (Chapter 34) or published outside the *Reference Frame* format (Chapter 35). Chapters 36–38 are hitherto unpublished expanded versions of the *Nobel Diary* (Chapter 19), *Elegance* (Chapter 22), and *Questions for 2100* (Chapter 24). Chapters 39–42 are in the manner of and complement "My life with Einstein" (Chapter 27), with the difference that I really did spend significant periods of my life with Michael Fisher, Walter Kohn, Ken Wilson, and Rudolf Peierls. Chapter 43 and much of Chapter 42 celebrate the importance of writing. Together they serve as my benediction to this whole collection.

I thank Simon Capelin for his understanding patience, and Leonid Levitov for getting me unstuck. Thanks also to Joan Feynman and Geoffrey Pullum for letting me reproduce their delightful letters.

Above all, I'm profoundly grateful to Gloria Lubkin for over two decades of wonderful collaboration.

PART ONE

Reference Frame Columns, *Physics Today* 1988–2009

1

What's wrong with this Lagrangean

A few months ago I found myself living one of my milder visions of hell, trapped on a flight to Los Angeles, having forgotten to bring along anything to read but *Physical Review Letters*. Finishing the two articles that had inspired me to stuff it in my briefcase before we even reached the Mississippi I decided to make the best of a bad thing by taking the opportunity to expand my horizons. Scanning the table of contents, I was arrested by a title containing the word "Lagrangean."

Funny, I thought, it's not often you see misprints so blatantly displayed. But when I turned to the article, there it was again, "Lagrangean," in the title and scattered through the text. Well, I thought, an uncharacteristic failure of the copyediting process. The authors were foreign and apparently didn't know how to spell. Copy editors aren't physicists, the word is surely in few if any dictionaries, and so it slipped through.

But I had nagging doubts. Easily resolved, I thought: You can't write an article in theoretical particle physics without a Lagrangian, so I can check it right now. Well, it turns out to be not quite that easy. To be sure, you can't do particle physics without a Lagrangian, but you don't have to call it anything more than L, and many don't. Nevertheless, I found a Lagrangian, fully denominated, in one more article, and there it was, shimmering derisively before my eyes again: "Lagrangean."

Now I am not a man of great self-confidence, and my secretary will testify that I am a rotten speller. Was I fooling myself? Could "Lagrangean" be right, and my conviction that it should be "Lagrangian" an orthographic hallucination induced by the absence of better things to read, like a mirage in the desert? Please ask yourself this, dear reader, before reading on: Would it have startled you?

When the plane landed in Los Angeles, I tooled up the freeway to the house of my hosts and breathlessly asked, "How do you spell 'Lagrangian'?"

"I dunno," said he, but she said without hesitation, "L-A-G-R-A-N-G-I-A-N," and I felt hope for my sanity, A quick tour revealed that every book in the house on mechanics and field theory spelled it with an *i*. I *was* sane! But what was going on at *Physical Review Letters*?

The next day at UCLA and the following day at Santa Barbara, I asked almost every physicist I met how to spell "Lagrangian" (all got it right) and whether they had ever noticed *Physical Review Letters* spelling it wrong (none had). A transitory anomaly, I thought—an accident limited to the issue I happened to put in my briefcase. But when I had a free moment I went off to the library, just to make sure.

This is what I discovered: *Physical Review Letters* has been systematically misspelling "Lagrangian" with an *e* instead of an *i* since the middle of 1985. At the start of July and earlier it is "Lagrangian"; by the end of the month and thereafter it is uniformly "Lagrangean." (In the interior of July 1985 it oscillates.) *They have been doing it for over two years*, and nobody I asked had noticed! Nobody I have asked since has noticed! Have *you* noticed?

The disease is confined to *Physical Review Letters*. As far as I can tell *Physical Review* in all its multitudinous varieties is still spelling the word correctly.

I am publishing this discovery here for the first time. I claim exclusive credit for it, my extensive random samplings having led me to conclude that nobody else ever noticed "Lagrangean" during the entire two and a half years it has been lurking in the pages of *Physical Review Letters*. My discovery raises at least two serious questions, of which I save the more disturbing for last.

Question 1. What is going on here? Why is *Physical Review Letters* misspelling "Lagrangian"? One can invent theories. To be sure, the man's name was Lagrange, ending, as any undergraduate can tell you, with an *e*. But if you write "Lagrangean," then shouldn't you pronounce it "luh-GRAN-jin," and doesn't everybody actually say "luh-GRAN-*jee*-in"? Doesn't "Lagrangean" lead unavoidably to "Hamiltonan," which gives me, for one, a case of the giggles, and certainly has never been sighted in the pages of *Physical Review Letters* or any other journal of repute? Ah, but "Hamilton" doesn't end with *e*. Well, what about people who do end with *e*? Try adjectivizing them. Don't you want to turn their *e* into an *i* before the *an*? Or do you?

Such talk, fascinating as it can become, utterly misses the point. English spelling is entirely irrational. Theorizing about it is a form of what Einstein called "brainschmaltz." There are no rules, only precedent. And precedent demonstrates unanimously, overwhelmingly, and unambiguously that for at least a quarter of a century the English word has been and remains "Lagrangian."

I devoutly hope the answer is that a bug crept into their spelling checker in the summer of 1985, but I fear the worst, and I therefore here declare that *Physical Review Letters* has no right to tamper with established usage. One can only hope the editors will soon come to their senses.

Question 2. The more disturbing question: Why has nobody noticed? Why did this aberration lie undiscovered for more than two years, only coming to light because one careless man allowed himself to fall into a dreadful trap that any prudent person would have taken simple measures to avoid? Can it be that physicists no longer know how to spell? No, because when I asked a random sample, they all spelled it "Lagrangian." Can it be that they are all speed readers, zooming on to the next word as soon as they get past the opening "Lag"? I don't think so. It seems to me that very fast readers take in whole lines at a time, and when you look at a whole line and you know how to spell, what you see glaring out at you, defiantly thumbing its nose, is "Lagrangean." We do know how to spell. We do see what we read. I can think of only one other explanation, but it is an explanation so alarming, so staggering in its implications, that I hesitate to give voice to it:

Can it be that nobody any longer reads *Physical Review Letters*?

We've known for some time that, roughly speaking, nobody any longer reads anything but preprints, the archival journal of choice, which for many years now has been *Physical Review Letters*, and secondary references cited in these two primary sources. But the preprints have been coming thicker and faster. And *Physical Review Letters* now publishes almost as many pages each month as *all* of *Physical Review* did back in 1956, when I was starting graduate school. (And at that time *Physical Review* included the letters as well as containing within one set of covers all of *A, B1, B15 I, B15 II, C, D1* and *D15*.) Yet slim as it was, and few as the other journals were, back in those easygoing days *Physical Review* was widely known as "the green plague." *Physical Review Letters* is now as big as the green plague of the 1950s, and the white plague (preprints) is even bigger. Is it then indeed possible that people have stopped reading it?

Few, of course, when asked about their reading habits will give you a straight answer, but I submit for public discussion what has to be regarded as a very powerful piece of evidence that the pages of *Physical Review Letters* are now examined no more than any of the other hundred thousand or so pages that pile up each month in our physical science libraries. I submit that whoever decided to start systematically misspelling "Lagrangian" was unwittingly (or could it have been wittingly?) conducting a beautiful experiment that could not have been more ingeniously contrived to get an honest measure of how carefully people actually look at *Physical Review Letters*.

The results of that experiment are disconcerting, with implications for at least two major problems that we have not adequately faced as a profession: the disaster looming over science libraries, and therefore over science itself, as a result of the irresponsible way we have allowed scientific journals to proliferate; and, not unrelatedly, the lamentable decline in the quality of scientific writing. I hope to address both of these problems in subsequent columns.

1. Postscript

1. Preprints are mentioned in this and several subsequent chapters. The term may be unfamiliar to younger readers, because they no longer exist, as prophetically suggested in Chapters 10 and 12. Before the internet became the dominant mode of communication, it was the practice of physicists to mail paper copies of papers submitted for publication to those professional colleagues, around the world, who they thought might be interested. Because these "preprints" arrived faster than any journals, they were read more assiduously than anything else, by those fortunate enough to receive them.

2. A letter to the editor of *Physics Today* (November 1988), commenting on *Lagrangean*, suggested that the time had come for major journals to cease paper publication in favor of electronic distribution. That vision turned out to be insufficiently radical. The writer predicted the journals would be shipped on disks to subscribing libraries.

3. I learned, just before *Lagrangean* appeared, that although he spent his career in France, Joseph Lagrange was born in Torino under the name of Giuseppe Ludovico Lagrangia, thereby undermining what little incentive there might have been to call his invention a Lagrangean. I reported this discovery in a line

added in proof. This elicited a letter from Turin indignantly informing me that "Lagrangia" was an abominable piece of fascist revisionism. But no evidence of its inauthenticity can be found from Google, a scholarly tool that was unavailable and unimaginable in 1988.

4. A year or two after *Lagrangean* appeared, *Physical Review Letters* quietly returned to spelling "Lagrangian" the same way as everybody else. I might be the only one who noticed.

2

What's wrong with this library

An extrapolation of its present rate of growth reveals that in the not too distant future Physical Review *will fill bookshelves at a speed exceeding that of light. This is not forbidden by relativity, since no information is being conveyed.*

I first heard this joke from Rudolf Peierls in 1961. Since then the size of *Physical Review* has doubled, doubled again, and is on the verge of completing its third doubling. *Physical Review Letters* is nearly as big as *Physical Review* was in 1961. The journals of other physical societies have undergone comparable expansions, as have the numbers and sizes of the commercially published physics journals.

I was recently asked to address a group of Cornell alumni and librarians on how the library is used by a typical physical scientist. I did point out that they had selected a distinctly oddball specimen, but they persisted. So to give them a sense of the calamitous conditions we have created, I went systematically through the current periodicals section of the Cornell Physical Sciences Library to count how many journals I felt I ought to look at but didn't. My criterion was stringent: a journal made my list only if it would be downright embarrassing to admit that I never looked. I had to be able to imagine a colleague responding to my confession with amazement: "You never look at X?!" Counting only once those with multiple versions (A, B, C,...) I still found 32. In one case, uncertain whether an admission that I never looked would occasion a blush or merely a nervous giggle, I randomly opened a random issue and found an article I had not seen before, expanding interestingly upon earlier work of my own. *That* information had not been conveyed.

It is in this context that we must view the resistance of many scientists to current efforts by library administrators to make draconian cuts in their subscription lists. Keeping up with everything is driving the libraries into bankruptcy. Early signs of this problem and how physicists might respond surfaced over ten years ago. My colleague Ken Wilson and I, physics department representatives on the library committee and therefore conscious of the looming catastrophe, responded to announcements of two new physics journals by writing a letter to *Physics Today*[1] announcing that our library could not afford any more subscriptions. We warned potential authors that if they published in the two newcomers they would therefore be unread at Cornell, and urged for the sake of science libraries everywhere that people should stop writing for, lending their names to the editorial boards of, and instructing their libraries to subscribe to these and other unnecessary new publications.

From around the world we were attacked with a fury I couldn't have imagined. Many held us to be the running dogs of Jim Krumhansl, also at Cornell and at that time editor of *Physical Review Letters*, in a blatant conspiracy to stifle all competition. Nobody believed for a minute that our concern was only, as we stated emphatically, for the survival of our library. As it turned out, one of the journals died within a year or two, for reasons unrelated to our kamikaze attack; the other still exists and, I'm ashamed to admit, the last time I checked Cornell was receiving it.

Now, 12 years later, things have worsened. Our physical sciences library has been given a budget we can live within only by cutting our already pruned subscriptions by another third, and cuts on this scale are not uncommon at other libraries. Scientists have two common responses to this threat to the integrity of their libraries:

1 March 1976, page 11.

▶ *A university is obliged to maintain its science libraries at the levels its scientists require because the library is included among the indirect costs of sponsored research for which the granting agencies reimburse the university.*

This argument plunges its unwary proponent into a snake pit of accounting tricks and countertricks, ringed about with ethical analysis that makes Kant read like Ann Landers. In its most coherent form the opening salvo goes like this: Journals are as important a part of doing research as magnets or liquid nitrogen. Rather than do without, we would include their cost in the budgets of our grants, but we cannot because library costs have already been taken into account in the indirect-cost calculation. The university has already been furnished the means to provide for our library needs. If those funds are insufficient the indirect-cost formula should be renegotiated. If they *are* sufficient ... (here the analysis flows into various vituperative channels that need concern us no further). (This argument is rapidly being rendered obsolete, at least in my own field of condensed-matter physics, by the increasing inability of the agencies to provide support even at the level of magnets and liquid nitrogen, but this seems not yet to have caught up with proponents of the Argument from Overhead.)

▶ *Alright [damn you], cut. But be sure to dovetail your cutting with other libraries in the city [state, region] so any journal can still be found within 25 [50, 200] miles from here and rushed over within 4 [8, 24] hours.*

This short-term fix will reduce the circulation of many journals, and we know how publishers with a captive audience will respond (and never were audiences more captive than university libraries to science journals): by raising their prices. Under this scheme each library will soon pay as much as before for its diminished collection.

Both the Argument from Overhead and the Dovetail Fix miss one central fact: There are too many journals. The problem is not how to persuade library administrations or granting agencies to produce funds to keep them all at our fingertips, nor is it to how to keep them readily available from a wider library pool. The problem is how to get rid of them.

It is a little hard to figure out how professionally rational people managed to land themselves in such a mess. Part of the blame must be assigned to the practice of good science libraries, now finally abandoned, of subscribing to everything regardless of cost, thereby offering publishers an irresistible incentive to launch new journals regardless of need. (One of the less attractive manifestations of high-temperature superconductivity is the demonstration it affords that the era of such launchings has not entirely died.) But why did we collaborate in this milking of our profession by sending publishers our articles and agreeing to be on their editorial boards whenever they saw another opportunity to rip us off? *Vita* enhancement is surely an insufficient motivation. In some cases it seems to have been nothing more than shortsighted penny-pinching. I have been told, for example, that it is impossible to do research in theoretical particle physics without a commercially published journal for whose two dozen neat little volumes we currently pay $4000 a year. Why do so many particle theorists publish there rather than in *Physical Review D*? You guessed it: no page charges.

We are probably stuck with that pricey little item for good, but most of the redundant journals have yet to reach the level of an addiction. Getting rid of them requires precisely the opposite of the dovetailing strategy. When my library decides to cancel a journal, far from it being important for neighboring libraries to retain it, they and most other libraries must cancel too. The journal will then expire because the few remaining subscribers, abused as they have been in the past, will not put up with a further massive increase in price, and because nobody wants to publish in a journal that significant numbers of their colleagues will not even have the opportunity to feel embarrassed at not examining.

How can we bring this about? To some extent it will happen automatically. As physicists across the country are asked by their libraries to specify what is to be dropped, provided we are not seduced by the dovetailing fallacy, the various lists of journals we can survive without will have many members in common, if there is any objectivity in our science. Harder to deal with will be those many journals with distinguished boards of editors in which 5 percent of the contents are respectable and even interesting. How are their numbers to be reduced?

Various schemes come to mind that I am not worldly enough to assess the feasibility of, much less bring to pass:

▶ A special evening session at every major APS meeting at which representatives of the library committees at universities and government and industrial laboratories would get together and agree on a common list of recommendations for the axe in that subfield.

▶ If the powers of the American Physical Society considered such a spectacle unedifying. what about providing a central clearing-house of information on recent cancellations? Libraries would inform the clearinghouse of their cancellations, and a periodic newsletter (*Look Who's Not Subscribing to What!*) would be sent to libraries and other interested parties, listing recent cancellations by journal and by library. To achieve the desired end it would probably suffice to circulate widely the cancellations of 30 or so major university libraries and a dozen or so industrial and government ones.

▶ If the APS will not participate even in this effort (and maybe it shouldn't, since, after all, it publishes journals too) then departmental library committees could do it informally by swapping lists of recent cancellations. "Thought you might like to know that last month we decided we could do without the following 37 and canceled. What have you been up to lately?"

Restraint on the part of individuals is also essential. The default response to a request to be on the editorial board of a new publication should be an emphatic no, accompanied by a reasoned statement of why the journal should not be started at all. You should review all journals on whose editorial boards you currently sit, asking whether the world would be significantly worse off if the papers appearing there were published elsewhere. In most cases you will have to admit that the world would on balance be better off. You must then resign immediately, explain why, and send your distinguished fellow editors copies of your resignation for their moral improvement.

Most importantly, however, and perhaps most painfully, we should all think twice before writing yet another article. For although significant economies can certainly be achieved by channeling the great flood of papers into fewer vessels, we (and the agencies that give us grants and the authorities that approve our promotions) have forgotten that the aim of publishing articles is to communicate. The current crisis of our libraries offers us a rare opportunity to draw back from our relentless march toward that dreaded point beyond which no information is conveyed.

2. Postscript

1. Since 1988, when this essay appeared, the number of pages of both *Physical Review* and *Physical Review Letters* has more than doubled. Not as bad as between 1961 and 1988, but not much help for one trying to keep up with the literature. But nobody even pretends to any more.

2. The most dramatic development in the next two decades was entirely unanticipated: the birth and growth of the internet, and

with it the online publication of journals. Online journals eliminated the hope that the size of scientific journals would soon be limited by the storage capacity of the libraries that held them.

3. Online journals have led to a phenomenon that had not occurred to me in 1988: the disappearance of libraries. At my own university the Cornell Physical Sciences Library, one of the finest of its kind in the world, was emptied of its contents and converted into a study hall in 2009. Those of us who had loved that library had a small wake in what used to be the current-periodicals room on its last day, with cheese, crackers, and wine. The library closed because nobody was going there any more to peruse journals. Everything was available online, either at home or in your office. Those of us who still had a quaint interest in books were assured that they would be distributed among the Mathematics, Engineering, and Biology Libraries, a short, healthy, and enlivening walk from the site of the former Physical Sciences Library. But a year later it was announced that the Engineering Library was also closing. As of 2015 Mathematics and Biology remain open.

If somebody had suggested these closures to me in 1988, I would have dismissed the idea as too implausible even to make good satire.

3

What's wrong with these prizes

But "glory" doesn't mean "a nice knock-down argument."

– Alice to Humpty Dumpty

It seems to me evident that the system of prizes, honors, and awards in physics has run completely amok, absorbing far too much of the time and energy of the community in proportion to the benefits conferred. Yet nobody complains. Every month *Physics Today* routinely announces the latest crop of winners, all the major American Physical Society meetings have sessions to bestow prizes, the APS directory continues to distinguish the asterisked from the unasterisked, and nobody ever complains. Why?

To ask the question is to answer it. Indeed, merely by publishing the above paragraph I have probably already irreparably blemished my reputation in the profession, and if *Physics Today* has actually printed this column I imagine it can only have been after heated and prolonged editorial debate. Much of this essay, in fact, sat aging in my computer in a directory with highly restricted access for almost two years. It was finally sprung loose by the 1988 presidential campaign, which filled me with so intense a loathing for those who hesitate to speak provocative truths that I can longer restrain myself. Here I go.

Why does nobody ever complain? Nobody complains because there are two categories of physicists: those who have won prizes and those who have not. Winners cannot criticize the system. It would be rude to the donors of their prizes. It would be offensive to the committee that selected them and the people who wrote letters on their behalf. It would be a vulgar display of bad taste. It would be unseemly to criticize a system one has benefited from before others have had their chance to win.

But neither can nonwinners criticize the system. It is not that a public attack on, for example, the absurdity of election to the National Academy of Sciences might jeopardize one's own chances for immortality, for this would be a noble sacrifice. What freezes dissent for the nonwinner is that it would be perceived as sour grapes—an unbecoming outburst of petty jealousy. The only respectable stance for the nonwinner is warmly to congratulate each new crop of winners, a kind and gentle response to be sure, but one that implicitly endorses the system itself, preposterous as it is.

At this point you may well be distracted from my original contention by the question of which camp you are being addressed from. I have wrestled at some length with whether to declare myself at the outset or force the curious into a possibly quite lengthy perusal of various arcane archives. The only satisfactory solution I have come up with is to invite anyone wanting to know to send me a stamped, addressed envelope, which I promise to return with an up-to-date CV.

Interestingly enough, by leaving unspecified my own level of glorification it seems to me that I am, at least with those readers who deem it as likely that I am glorified as not, doing considerably less damage to my reputation for courtesy, tact, and simple decency than I would have done had I declared myself explicitly to be either of the two (exhaustive and mutually exclusive) types. This is as close to a demonstration of quantum interference on the sociological level as I have ever encountered. But I digress.

I realized that the honor system had become a destructive force shortly after having assumed certain administrative responsibilities. Before that I had never thought much about it one way or the other, occasionally submitting essays on behalf of deserving people I thought had been overlooked, noting with pleasure the good awards, and with irritation or amusement, the bad ones. Only recently did I learn ("How innocent can you get!" you will say, dear reader—you who have known the dark side of awards longer than I and yet have never spoken up publicly against the whole business) that these things are systematically sought after by organized campaigns, routinely consuming oceans of time and effort.

If we don't put up all of our guys, they'll win with theirs, seems to be the guiding principle. No point in disinterestedly recommending the most deserving, irrespective of institutional affiliation, for such people are already being backed by their own teams. Conversely, if we don't push our own, nobody else will. The folklore in my corner of physics is that it's the industrial laboratories that put up the most massive and systematic campaigns, but in my experience the universities have been quick to acquire the bad habits of all whom they deal with, and I wouldn't want to say who are the worst offenders.

Once you start down this path the process acquires a crazy momentum. If you have put across a winner you can't sit back and enjoy the satisfaction of a job well done. Can one rest after X gets prize A? Certainly not: 65% of all winners of the A Prize go on to receive the B Medal, half of the B Medalists become fellows of the DE of F, and it would be an irresponsible administrator who didn't go for the whole pile. Worse, as even the slightest aura of glory becomes attached to routine professional activities—for example, giving a talk at a meeting—the point of selecting people for such jobs flips from finding the best to supporting the team [which in the case of my team (but not yours) amounts to exactly the same thing].

This stampede after glory, foreign and domestic, would be a piece of harmless silliness, did it not involve such a substantial expenditure of human energy. Most of us are asked to make other judgments that, unlike the cosmetic decisions in the glory game, are of vital importance to the professional survival of our colleagues. We are asked to review grant proposals, we are asked to referee manuscripts, we are asked to evaluate colleagues for appointments to new positions or promotions. No responsible member of the profession could refuse to do these things, but most of us do so many of them that we don't do a very good job. We simply haven't the time.

I maintain that with all these serious demands on our attention, this childish scramble after glory is a frivolity we can no longer afford. How to relieve ourselves of it is less clear. It would be too much to hope for the abolition of all prizes and self-perpetuating honorary societies. The child in each of us cannot, and probably should not, be entirely obliterated. Baseball understands these things, and does them much better than we do, conferring the Most Valuable Player Award by decision of the sportswriters, leaving the players themselves to get on with more serious business. A moment's reflection on the spectacle of even the top science writers voting to select, say, the Physics Rookie of the Year reveals that this won't work. As a spectator sport, physics is a complete bust. The rules are too complicated, and the science writers can't really judge performance.

My guess is that it is up to the people who make these distinctions to save the rest of us from this frenzy of unproductive effort. It would be unfair to ask selection committees to refuse all external nominations and do the entire job themselves, though whatever else one might think about the MacArthur Awards, they do have the not inconsiderable virtue of wasting the time of relatively few in the selection process. But could not the bestowers of prizes limit to one the number of people they were willing to hear from in support of any given momination? Suppose it were specified that there would be a preliminary screening of all letters nominating a candidate to determine which single one was to be retained, the others being destroyed without any record of them or their authors. Presumably any qualified observer can summarize the nominee's accomplishments. Stirring up the mighty of the Earth to bombard committees with letters of their own repeating these data is a ritual it is high time to set aside. Let the committee make a few phone calls if it wants confirmation of the one letter.

Better yet, why can't people nominate themselves? Indeed, why not insist on it? Who, after all, is better qualified to prepare the case, and more likely to do it with verve and enthusiasm? We are already, with only a few unfortunate exceptions, the only ones who nominate ourselves for research grants or our own prose for inclusion in prestigious journals. Should I die wealthy I will endow an APS prize (probably for Theoretical Contributions to Statistical and Low-Temperature Physics by One with a Fine Prose Style). The Mermin Prize will be available only to applicants who submit an essay of no more than 500 words demonstrating explicitly and implicitly why they qualify, a list of no more than eight relevant papers, and the names of two people the committee might or might not want to consult in a phone conversation of no more than three minutes' duration. The names of all applicants will be published in the APS *Bulletin* to discourage the frivolous and install a proper humility in the serious, for the point of the Mermin Prize will be not glory but money—$750 000 sounds good. Applications will remain valid for four years, no updating permitted, after which unsuccessful applicants will become ineligible. People with no interest in the process can go on peacefully doing physics.

I offer these views in the hope that having thus shot myself in the foot, I may encourage others to voice their opinions on what, if anything (hold on for a breathtaking swerve of metaphor), this particular emperor is or ought to be wearing. Can't we discuss this business out in the open? Or is it too much like explaining on prime-time television that it's *wrong*—never mind unconstitutional—to force people to pledge allegiance to a flag?

3. Postscript

1. Nobody took up my invitation to send them a CV. Note my quaint suggestion that readers interested in this information should send me a stamped, addressed envelope.

2. I said I was publishing my critique of prizes because I was disgusted by the reluctance of the participants in the 1988 U.S. Presidential election to speak provocative truths. The final sentence of the column alludes to an example in which George Bush the First pilloried his opponent Michael Dukakis for defending the constitutional right of a religious group to refuse to pledge allegiance to the flag of the United States of America.[1]

   But the real reason why I published the column when I did is that I received a phone call informing me that I had just been awarded a prize by the American Physical Society. I realized that when this information was made public, publishing my column would indeed be rude to the donor of the prize and offensive to the committee that selected me. So I wanted to get the column into print as soon as possible. It appeared just as the prize was announced and a little before the meeting of the APS at which the prize was presented. The official presenting the prize joked that he was surprised I had accepted it. But it was $12,000 (1989 dollars!) and my daughter was about to start college, so this was not a difficult decision.

3. My remark about the absurdity of election to the National Academy of Sciences elicited an unexpected reaction from one of its members. Serge Lang, a Yale mathematician (who wrote beautifully about mathematics for the general reader), was

---

1 How times do change! In 2015 the American right wing has taken to insisting that religious groups have a constitutional right to ignore the constitutional rights of others.

elected to the Academy a few years before my column appeared. The primary activity of most members of the Academy is to participate in the interminable year-long process of electing more members. This is made difficult by the requirement that in the final round of balloting, in order to have your vote counted in the area you know something about, you must also vote for a specified minimum number of candidates in all areas of the Academy, in most of which you have no expertise whatever, and therefore no ability to make well-informed judgments.

Lang was one of the few members of the Academy to take this responsibility seriously, studying the works of many nominees in fields quite remote from his own. Occasionally he came upon people in distant fields who struck him as profoundly unqualified. He then felt it his duty to warn the general membership about their proposed election, and he developed quite extensive mailing lists for this purpose.

Shortly after my column appeared Lang sent an enthusiastic notice to his membership list under a heading something like "Mermin defies Academy!" Having long admired his writings about mathematics, I basked in his approval. Since I had never considered myself suitable material for Olympus, his celebratory underlining of my dismissive remark didn't disturb me.

But, strangely enough, two years after my column appeared I was elected to the Academy. I think Lang's publicizing my disparaging remark actually helped me. Because all members are required to vote for a specified minimum number of nominees in all areas, no matter how unqualified they are to make that judgment, name recognition is important. If you've heard of somebody in a remote field you're more likely to vote for them. Absurd indeed!

4

What's wrong with this pillow

Attitudes toward quantum mechanics differ interestingly from one generation of physicists to the next. The first generation are the founding fathers, who struggled through the welter of confusing and self-contradictory constructions to emerge with the modern theory of the atomic world and supply it with the "Copenhagen interpretation." On the whole they seem to have taken the view that while the theory is extraordinarily strange (Bohr is said to have remarked that if it didn't make you dizzy then you didn't really understand it), the strangeness arises out of some deeply ingrained but invalid modes of thought. Once these are recognized and abandoned the theory makes sense in a perfectly straightforward way. The word "irrational," which appears frequently in Bohr's early writings about the quantum theory, is almost entirely absent from his later essays.

The second generation, those who were students of the founding fathers in the early postrevolutionary period, seem firmly—at times even ferociously—committed to the position that there is really nothing peculiar about the quantum world at all. Far from making *bons mots* about dizziness, or the opposite of deep truths being deep truths, they appear to go out of their way to make quantum mechanics sound as boringly ordinary as possible.

The third generation—mine—were born a decade or so after the revolution and learned about the quantum as kids from popular books like George Gamow's. We seem to be much more relaxed about it than the other two. Few of us brood about what it all means, any more than we worry about how to define mass or time when we use classical mechanics. In contemplative moments some of us think the theory is wonderfully strange and others think it isn't; but we don't hold these views with great passion. Most of us, in fact, feel irritated, bored, or downright uncomfortable when asked to articulate what we *really* think about quantum mechanics.

I'm one of the uncomfortable ones. If I were forced to sum up in one sentence what the Copenhagen interpretation says to me, it would be "Shut up and calculate!" But I won't shut up. I would rather celebrate the strangeness of quantum theory than deny it, because I believe it still has interesting things to teach us about how certain powerful but flawed verbal and mental tools we once took for granted continue to infect our thinking in subtly hidden ways. I don't think anybody, even Bohr, has done an adequate job of extracting these lessons. From this point of view the problem with the second generation's ironfistedly soothing attitude is that by striving to make quantum mechanics appear so ordinary, so sedately practical, so benignly humdrum, they deprive us of the stimulus for exploring some very intriguing questions about the limitations in how we think and how we are capable of apprehending the world.

I would guess that an unvoiced reason for such efforts to render quantum mechanics uninterestingly bland is the desire to counter the kind of dumb postquantum anti-intellectualism that says that even the physicists now know that everything is uncertain, leading to the disastrous corollary: Anything goes. It is indeed important to emphasize to those who would go from quantum mechanics to know-nothingism that the quantum theory, far from filling us with paralyzing (or liberating) uncertainty, now permits us to make the most accurate quantitative calculations in the history of science. We must certainly speak up against "the general antirationalist atmosphere which has become a major menace of our time, and which to combat is the duty of every thinker who cares for the traditions of our civilization."

On the other hand it's important in combat to shoot at the right target. The above quotation is from Karl Popper and is directed against the writings of Heisenberg and Bohr. Physicists in the second generation certainly have a much better sense of where to direct their fire; but in sanitizing the quantum theory to the point that nothing remarkable sticks out above the surface you run the risk that if you go inside and look around you won't find anything left to make it hang together anymore.

Thus although it is a fact about the quantum theory of paramount importance that it permits us to calculate measurable quantities with unprecedented precision, it does not follow from this that statements that the quantum theory is not deterministic but acausal are vast exaggerations—that the theory has little to do with whether or not nature is a game of probability. Yet it has been argued in this context [1] that even radioactive decay—the very paradigm of acausal discontinuous quantum behavior—appears as probabilistic and abrupt only when an inappropriate question is asked: If a particle is in a state of very well-defined energy, then it is inappropriate to ask for the exact time of its decay, and the answer is probabilistic only because the question is not appropriate to the experimental situation.

Now to be sure I can, at least in principle, produce at noon a particle that will decay as the clock is striking midnight, provided I make it in some tricky superposition of energy eigenstates for which asking when the particle decays *is* the appropriate question. But that does not mean that the acausality and discontinuity I associate with the beta decay of a free neutron are somehow my fault, stemming from my having asked the wrong question. The argument that the decay is causal and smooth relies on the fact that the quantum state—which incorporates all there is to know about the neutron—changes continuously, without any jumps, and indeed deterministically, according to Schrödinger's equation. That's fine. Nevertheless, if I put the neutron into a spherical cavity lined with counters, there will be a rather well-defined "ping!" at a rather well-defined but unpredictable moment. "All there is to know about the neutron" may well be evolving continuously and deterministically, but that little guy in the cavity goes off discontinuously and probabilistically.

Something interestingly puzzling gets lost by insisting that we are confronted with discontinuity and probability only when we ask a foolish question. The puzzle has to do with the nature of the quantum state, and whether it should be viewed as describing the system, or as describing our knowledge of the system, or as some combination of both, or as none of the above because the quantum state is actually nothing more than an ingredient in a mathematical algorithm for computing the results of a well-defined experiment. Using the continuous and deterministic evolution of the quantum state to argue against discontinuity and indeterminism in the atomic world makes more or less sense depending on which of these positions you adopt. If indeed it is nothing more than "all there is to know about the system" that changes continuously and deterministically, then this says nothing about whether the world itself can change discontinuously and probabilistically, unless one takes the position that physics is not about the world but only about "all there is to know" about the world, to which I would say: "Ping! Thus do I refute you."

If, however, the state describes the system and not just our knowledge of the system, then I somehow have to think of a neutron as continuously and deterministically leaking electron, albeit in a six-dimensional configuration space (nine, if you count the antineutrino too). This introduces continuity and determinism. But the "ping!" is still there, now being induced by the interaction with the surrounding counters. Are the counters asking the wrong question?

Another thing frequently declared [2] by members of the second generation to be resolved by refraining from asking foolish questions is the puzzlement engendered in some by contemplating the Einstein–Podolsky–Rosen experiment. Usually what is offered in support of this claim is the observation that there is nothing mysterious in the measurement of the spin of one particle being correlated with the probability distribution of the spin of the other, even if the two are far apart, since the two particles originate from a common source. Nobody would quarrel with that, but what many people find mysterious is not the existence of such correlations but their particular character, which turns out to be utterly inconsistent with some extremely simple and apparently very reasonable ideas about the kinds of correlations it is possible to have between far-apart noninteracting systems exclusively as the result of their having once been together in the same place. It may well be that to ask for any explanation of this "unreasonable" character of the correlations is to ask a foolish question. But the question cannot fairly be dismissed as foolish without saying what it is and making explicit the simple and apparently reasonable ideas that have to be thrown out with it.

My own view of EPR, which keeps changing—I offer this month's version—is that barring some unexpected and entirely revolutionary new developments, it is indeed a foolish question to demand an explanation for the correlations beyond that offered by the quantum theory. This explanation states that they are the way they are because that's what the calculation gives. Some explanations may sound more profound than this—saying, for example, that the correlations are a simple consequence of angular momentum conservation—but that is only because they go into a little more detail about what goes into the calculation. There is, however, an interesting nonfoolish question: Why do many knowledgeable and thoughtful people feel so strongly impelled to ask the foolish one?

My current version of the answer, not very well developed, is that it has something to do with certain deterministic presuppositions that are built into our thought and language at some deep and not very accessible level, and that have somehow infected even the way we think about probability distributions. Being of this frame of mind, I am therefore unwilling to be told both that the importance of indeterminism in quantum mechanics has been grossly exaggerated *and* that there is nothing peculiar about the EPR correlations. Einstein once wrote to Schrödinger that "The

Heisenberg–Bohr tranquilizing philosophy—or religion?—is so delicately contrived that, for the time being, it provides a gentle pillow for the true believer." When I rest my head on a quantum pillow I would like it to be fat and firm; the recently available pillows have been a little too lumpy to soothe me back to sleep.

References

1. H. Feshbach, V. F. Weisskopf, *Physics Today*, October 1988, p. 9.
2. See, for example, reference 1 or F. Rohrlich, *Science* 221, 1251 (1983).

4. Postscript

1. My phrase "Shut up and calculate" passed into the culture of quantum mechanics. As noted in Chapter 26, it has been widely misattributed to Richard P. Feynman. I believe *Pillow* is the first time the phrase was ever applied to an interpretation of quantum mechanics. In Chapter 26 I take a kinder, gentler view of the Copenhagen interpretation, about which I have more to say in Chapter 33.

2. As a further indication of how vexing the interpretation of quantum mechanics can be, in rereading *Pillow* I find that I no longer agree with my brisk dismissal of the view "that physics is not about the world but only about 'all there is to know' about the world." "Ping! Thus do I refute you" is not much of an argument. My views in 2015 can be found in Chapters 30–33.

3. To my delight, I discovered from an ad in the *New York Times* (June 24, 1985) that there actually was a new brand of pillow called "quantum." I couldn't work this wonderful coincidence into my column, but I say more about the ad in Chapter 27. Today all kinds of things are called "quantum," but the word had hardly entered the vocabulary of commerce in the late 1980s.

5

What's wrong with this prose

I write bleary-eyed and disheartened, after a long proofreading session mainly devoted to inserting into the galleys calls for the restoration of what was capriciously and destructively altered in the editorial offices of *Physical Review*. I proofread simply by reading the galleys, without reference to the original manuscript. My writing is a process that does not converge: I cannot read a page of my own prose without wanting to improve it. Therefore when I read proofs I entirely ignore the manuscript except to check purely technical points. Proofreading offers one more shot at elusive perfection. Proceeding in this way. I come to the end of a paragraph with a lurching sensation. The last sentence seems to be a *non sequitur*. Can I be failing to get my own point? Turning to the copy-edited manuscript, I find a marginal message: "Author: Please note that we discourage single-sentence paragraphs." As an application of this principle, one short emphatic paragraph has been attached to the end of another, to which it is entirely unrelated. If you set asunder what *Physical Review* has joined, it makes sense again.

What is the justification for such a rule? Excessive use of single-sentence paragraphs blurs the distinction between the sentence and the paragraph, makes for a visually unattractive page and becomes boring. But the occasional single-sentence paragraph is a powerful device. It gives a pause in the rush of thought, it focuses attention, and it can contribute powerfully to the rhythm of the prose. The Constitution of the United States of America, whose prose Warren Burger enjoined us to admire in its 200th-anniversary year, is chock-full of beautiful single-sentence paragraphs. A blanket prohibition is absurd, and enforcing it by paragraph grafting is almost certain to do violence to the clarity and even the meaning of a well-written essay. So I go through the galleys restoring the three or four indigestibly merged paragraphs, adding my own marginal messages ("Editor: We discourage gratuitous confusion") in the hope that my counterinstructions will not be ignored.

A bit later I come to a reference to nature, "Nature herself," I remember writing, "has proved to be quite unambiguous…" The galley reads, "Nature has proven quite unambiguous…." Not bad, I think, getting rid of that unnecessary "to be"—should have spotted it myself. But then I notice that nature has been depersonified. Why can't nature be "she"? Could "herself" have been sacrificed in an enlightened attempt to exorcise unconscious sexism from the pages of *Physical Review*? No. (*Author: Please note that we discourage one-word sentences.*) The desexing of Mother Nature is explained by "Author: Please note that the editor feels this wording to be more literal, and therefore preferable." The note refers me to other applications of the same rule to my manuscript: The adjective has been deleted from a reference to a "charming monograph," and "aficionados of ring theory" has become "ring theorists." The first alteration has deprived the reader of the information that the work in question is uncharacteristically readable for a monograph on number theory; the second eliminates the information that ring theory is not part of the everyday mathematical equipment of most physicists, and also introduces an absurdly inappropriate pomposity (compare "evolution theorists" or "relativity theorists" or "group theorists").

The next thing I run into is "Author, please place only a word or short phrase rather than a whole sentence in italics." Well. OK, I can see that whole sentences in italics might make for a blotchy kind of page, particularly if there are lots of equations around. But occasionally it can be quite useful to call attention to a central point by putting it in italics. I maintain that anything you can do to help the reader follow your argument is worth doing. Nevertheless, I'm willing to forgo excessive use of the italic option for the sake of a neater page. But what have they done at *Physical Review*? They haven't removed *all* the italics; selected (God knows how) words in those formerly italicized sentences have been left in italics, with almost uniformly preposterous results, (My proofs sport about ten such *sentences*, all *reading* like this one; I freely *admit* that I probably got *carried* away with italicized sentences, but surely the cure is worse than the *disease*.)

Stranger still, in the caption of a geometric figure the assertion that the straight line joining B to point F has the same length as the straight line joining point A to point F, which appears in the manuscript as "$BF = AF$," has been transformed into "$d(BF) = d(AF)$ (where d is the distance)." This violates three cardinal rules at once: Do not introduce unnecessary notational complexity; do not introduce unnecessarily unconventional notation; do not make lengthy that which is brief.

And so it goes. *Physical Review* is certainly not the only practitioner of destructive copy editing. *Scientific American* is notorious for elephant-walking over the writing that enters its offices, systematically pounding it into homogeneous soporific mush. Even *Physics Today*, which publishes some of the better prose in the scientific literature, is not without its foibles. (I am told they are thinking of reforming.) Were you, for example, reviewing a concert for *Physics Today* you would be required to talk about "Wolfgang Mozart's *Jupiter* symphony" because, I can only imagine, the reader might be under the impression that Leopold wrote one too. It would have to be Johann von Goethe's *Faust* and Sandro Botticelli's *Primavera*. Everybody you mention in *Physics Today* has to have a first name. This is absurd and can also be destructive of good writing, introducing the literary equivalent of a hiccup into a smooth sentence, or raising in the reader's mind such spurious questions as "Why *Werner* Heisenberg; was there another I didn't know about?"

Why am I telling you all this? Surely you all have stories of assaults on your manuscripts as irritating as mine. Precisely. I raise the matter to urge you to fight back. This savaging of our prose—this obliteration of our human individuality—has something important to do with one of the great failures of science in our time: the virtual disappearance of just plain readable—never mind humane—scientific prose. This is a calamity for science, and not only because it makes the practice of science much less fun. Bad thinking is vastly easier to cover up if you're allowed to get away with and even encouraged to produce bad writing.

Among the principles underlying these examples of copy editing is the intention to eliminate any trace of a human author. The inevitable result is a bland uniformity. By making the point that anything remotely lively, idiosyncratic or quirky will be eliminated, *Physical Review* deprives an author of any incentive to write interestingly and, worse, makes it very much more difficult for an author to provide gracefully the kinds of emphases and signposts without which scientific exposition can become virtually unintelligible.

Eliminating the artificial obstacles to decent scientific prose erected by *Physical Review* will not in itself ensure the return of clean and vibrant writing to its pages, but as long as the copy-editing process continues to emasculate or defeminize our texts, there is no hope that we can breathe the life back into scientific writing or persuade our students that writing well is a worthy and even noble endeavor. The final result of our efforts as scientists is, after all, not a table of data, a set of equations, or the output of a computer. It is an essay, a piece of expository prose. That's what grant officers, promotion committees, and biographers care about and for once they're right.

So fight back. Restore the humanity to your bowdlerized text when the galleys arrive. Victory does not come easily, but it will never come to those who refuse to fight. I changed "monograph" back to "charming monograph" in the proofs, and I write now, over a year since I began this essay, to report the results. I got a call from *Physical Review*.

"About that monograph…," the man said.

"Yes?"

"How would you like 'interesting monograph' or 'important monograph'?"

"Well," I said, "as a matter of fact it isn't *terribly* interesting. And *nobody* could honestly say that it was important." Long pause. "But it *is* charming."

"Oh," he said. 'I see."

And it stayed "charming." Therefore do not hesitate to write interesting, readable, lively, intelligible articles. It is your duty to do so. And when the proofs come back duller, clumsier, and more ambiguous than the manuscript you sent in, restore the life to those galleys, and be calm but firm when the phone rings. You will not only have more fun that way, but you will also be contributing to the good fight to reverse the sad and dangerous decline of scientific discourse in our time.

5. Postscript

1. Several years after *Prose* appeared I sent a paper to *Physical Review Letters* with the title "Beware of two-dimensional lattices with 46-fold symmetry." On the form acknowledging receipt of the manuscript was added in handwriting: "We question the suitability of the title. It is catchy but doesn't convey much information."

Had the title been (as it was in an earlier draft) "Uniqueness of two-dimensional lattices with n-fold symmetry for $n < 46$", its suitability would not have been questioned, even though it leaves open the question of whether we stopped at 46 because 45 is already quite large enough, because it just got too hard to press the frontier further at that point, or because 46 is truly exceptional. The word "beware" aptly conveys both the complacency appropriate to cases less than 46, and the trouble this can land you in if you blithely assume that 46 will be no different from the earlier cases.

What actually disturbed the editor was the title's invocation of raw human emotion. Titles of physics papers rarely address the feelings their contents ought to inspire in the reader. The dominant tradition in late 20th- and early 21st-century scientific prose has been to produce something suitable for direct transmission from one computer to another, from which any trace of human origin has been purged, and in which any suggestion of the humanity of the author or the reader would be in bad taste.

After the paper had been reviewed, I received from the editors an unusual letter of acceptance: "We have decided to publish the paper provided you submit an appropriate informative title. We find that your title is catchy, but not informative. We don't mind catchy titles if they are also informative." I sent them an indisputably informative title that continued to begin with "beware". They accepted it.

2. *Physical Review* once had a rule against the first person singular. The editorial "we" was mandatory. Since most physics papers have multiple authors the issue did not often arise, but I very much like to write papers by myself. It is not just pompous to make "we" the authorial voice in a single-author paper. It deprives you of an opportunity to distinguish gracefully between when you're speaking for yourself and when you have in mind both yourself and your reader. "I [the author] emphasize that with this approach we [any of us] can rapidly solve the problem."

With such examples, I persuaded *Physical Review* to allow me to use the first-person singular. This lasted for a year or two. Then they made a new rule that single-author papers *must* use the first-person singular. "We" was prohibited unless there were multiple authors. After a similar conversation, they relaxed the rule to readmit the first-person plural.

6

What's wrong with these equations

major impediment to writing physics gracefully comes from
the need to embed in the prose many large pieces of raw mathematics. Nothing in freshman composition courses prepares us for
the literary problems raised by the use of displayed equations. Our
knowledge is acquired implicitly by reading textbooks and articles,
most of whose authors have also given the problem no thought.
When I was a graduate teaching assistant in a physics course for
nonscientists, I was struck by the exceptional clumsiness with
which extremely literate students who lacked the exposure even to
such dubious examples treated mathematics in their term papers.
The equations stood out like droppings on a well-manicured lawn.
They were invariably introduced by the word "equation," as in
"Pondering the problem of motion, Newton came to the realization that the key lay in the equation

$$F = ma". \tag{1}$$

To these innocents equations were objects, gingerly to be pointed
at or poked, not inseparably integrated into the surrounding prose.

Clearly people are not born knowing how to write mathematics. The implicit tradition that has taught us what we do know
contains both good strands and bad. One of my defects of character being a preference for form over substance, I have worried
about this over the years, collecting principles that ought to govern the marriage of equations to readable prose. I present a few of
them here, emphasizing that the list makes no claim to be complete. We are constantly assaulted by so many egregious violations

of even these simple precepts that I offer them in the hope that a few sinners—not only writers, but copy editors, publishers of journals, and even the authors of the mathematics subsections of literary style manuals—may read them and repent the error of their ways, or even be inspired to further beneficial studies of the sadly neglected field of mathematico-grammatics.

Rule 1 (Fisher's rule). This rule, named after the savant who reprimanded me for abusing it when I was young and foolish, simply enjoins one to *number all displayed equations*. The most common violation of Fisher's rule is the misguided practice of numbering only those displayed equations to which the text subsequently refers back. I call this heresy Occam's rule. Back in the days of pens, pencils, and typewriters, use of Occam's rule was kept under control by the pain of having to renumber everything by hand whenever it was deemed wise to add a reference to a hitherto-unremarked-upon equation. One often encountered papers displaying the results of the ungainly Fisherian–Occamite compromise: Number all displayed equations that you think you *might* want to refer to. Now that automatic equation-numbering macros can act upon symbolic names, the barrier to full Occamism has been removed, and it is necessary to state emphatically that Fisher's rule is for the benefit not of the author, but the reader.

For although you, dear author, may have no need to refer in your text to the equations you therefore left unnumbered, it is presumptuous to assume the same disposition in your readers. And though you may well have acquired the solipsistic habit of writing under the assumption that you will have no readers at all, you are wrong. There is always the referee. The referee may desire to make reference to equations that you did not. Beyond that, should fortune smile upon you and others actually have occasion to mention your analysis in papers of their own, they will not think the better of you for forcing them into such locutions as "the second equation after (13.21)" or "the third unnumbered equation from the top in the left-hand column on p. 2485." Even should you solipsistically choose to publish in a journal both unrefereed and unread, you might subsequently desire (just for the record) to publish an erratum, the graceful flow of which could only be ensured if you had adhered to Fisher's rule in your original manuscript.

Rule 2 (Good Samaritan rule). A Good Samaritan is compassionate and helpful to one in distress, and there is nothing more distressing than having to hunt your way back in a manuscript in search of Eq. (12.47) not because your subsequent progress requires you to inspect it in detail, but merely to find out what it is *about* so you may know the principles that go into the construction of Eq. (7.38). The Good Samaritan rule says: *When referring to an equation identify it by a phrase as well as a number.* No compassionate and helpful person would herald the arrival of Eq. (7.38) by saying "inserting (2.47) and (3.51) into (5.13)..." when it is possible to say "inserting the form (2.47) of the electric field **E** and the Lindhard form (3.51) of the dielectric function ε into the constitutive equation (5.13) ...." To be sure, it's longer this way. Consistent use of the Good Samaritan rule might well increase the length of your paper by a few percent. But admit it. Your paper is probably already too long by at least 30% because you were in such a rush to get it out that you didn't really take enough care putting it all together. So prune elsewhere, but don't force your poor readers—you really *must* assume you will have some, or it is madness to go on writing—to go leafing back when a few words from you would save them the trouble.

Admittedly, sometimes an equation is buried so deep in the guts of an argument, so contingent on context, so ungainly in form that no brief phrase can convey to a reader even a glimmer of what it is about, and anybody wanting to know why it was invoked a dozen pages further on cannot do better than to wander back along the trail and gaze at the equation itself, all glowering and menacing in its lair. Even here the mere attempt to apply the Good Samaritan rule can have its benefits. If the nature of the equation is inherently uncharacterizable in a compact phrase, is the cross-reference really necessary? Indeed, is the equation itself essential? Or is it the kind of nasty and fundamentally uninteresting intermediate step that readers would either skip over or, if seriously interested, work out for themselves, in neither case needing to have it appear in your text? If so, drop it. You will then have to revise the argument that referred back to it, but the chances are good that the argument will gain in clarity from not having an uncharacterizable monster of an equation at its heart.

Rule 3 (Math Is Prose rule). The Math Is Prose rule simply says: *End a displayed equation with a punctuation mark.* It is implicit in this statement that the absence of a punctuation mark is itself a degenerate form of punctuation that, like periods, commas, or semicolons, can be used *provided it makes sense*. For unlike the figures and tables in your article, unlike droppings on a lawn, the equations you display are embedded in your prose and constitute an inseparable part of it. The detailed theory of how equations are to be viewed as prose need not concern us here. Sometimes they function as subordinate clauses, the equals sign being the verb; sometimes they appear as substantive phrases, like a list of the contents of a room; sometimes, regrettably, they must merely be presented to the reader as objects like quotations (but with the convention that quotation marks are not required [except in the rare case that Math Is Prose requires it, as in Eq. (1) above (which I never dreamed I would be referring back to when I first put it into this essay)]).

Regardless, however, of the often subtle question of how to parse the equation internally, certain things are clear to anyone who understands the equation and the prose in which it is embedded. Thus the end of the equation may or may not coincide with the end of the sentence in which it stands. If it does, then the equation should end with a period or, rarely, if the equation terminates an interrogative sentence, it should end with a question mark. (Having now succeeded in publishing an equation requiring a quotation mark, it remains my dream to publish an article with an equation that requires a question mark; somehow I haven't got around to it.) If the equation terminates a clause or is part of an extended list, then it should end with a comma or semicolon. Only infrequently is no punctuation required, as, for example, in "Only when

$$\sum_{i=1}^{N} f(x_i) = 0 \tag{2}$$

is it impermissible to divide by this sum."

We punctuate equations because they are a form of prose (they can, after all, be read aloud as a sequence of words) and are therefore subject to the same rules as any other prose. To decree that every sentence should end in a period *unless* the sentence terminates in a displayed equation is grotesque. (If you disagree, try the rule that every opening quotation mark must be followed by a closing one unless the quotation terminates in an equation.) But one does not punctuate equations only because it is ugly not to; more importantly, punctuation makes them easier to read and often clarifies the discussion in which they occur. Acquiring the habit of viewing an equation not as a grammatically irrelevant blob, but as a part of the text fully deserving of punctuation, can only improve the fluency and grace of one's expository mathematical prose.

Most journals punctuate their equations, even if the author of the manuscript did not, but a sorry few don't, removing all vestiges of the punctuation carefully supplied by the author. This unavoidably weakens the coupling between the math and the prose, and often introduces ambiguity and confusion. I'm sorry to say that *Physics Today* is guilty of this practice. To be sure, its use of equations is sufficiently light that this does not inflict substantial hardship on readers, but it greatly undermines the role *Physics Today* so commendably plays in other respects as a model of good writing about hard science. May the appearance of Eq. (1) above signal the start of a new and better tradition.

We should strive, more generally, to make errant journals mend their ways. It is easier than you might think. One of my students and I once did a piece of work that required us to lead the reader (or at least, we know for a fact, the referee) through unavoidably dense thickets of equations. Unfortunately the otherwise obvious journal for our paper systematically violated the Math Is Prose rule, so in our letter of submission we emphasized that the punctuation in our equations was essential for the comprehensibility of our argument. The letter of acceptance, however, informed us that the publisher adhered in this and all its other journals, as well as in its books, to a firm policy of never punctuating equations. In that case, we wrote back, just return the manuscript and we'll send it somewhere else. After a long pause we were informed that at a meeting of the board of directors of the publishing firm a special dispensation had been granted to our paper, and indeed, it appeared with punctuated equations [1].

Fortunately Fisher's rule and the Good Samaritan rule don't require assent from boards of directors, so you have nobody to blame but yourself if your papers don't observe them; you can mend your ways right now. At a minimum you will make life much easier for an overworked referee, and with luck you might even have a few happily undistressed readers.

Reference

1. A. Garg, N. D. Mermin, *Foundations of Physics* 14, 1 (1984).

6. Postscript

1. I learned "Fisher's Rule" from my friend and former Cornell colleague Michael E. Fisher, who spent twenty-five glorious years at Cornell as the Horace White Professor of Chemistry, Physics, *and* Mathematics, during which he taught me many important things. In Chapter 39 I describe more of these lessons.

2. *Equations* elicited several letters to the editor of *Physics Today* (June 1990). One writer argued strenuously against Fisher's Rule, maintaining that equation numbers were themselves an important kind of punctuation, identifying the important ones for the reader. In my reply I asked whether he would

enjoy looking something up in a book that, eschewing the mindless convention that all pages have numbers, reserved them for only the really important ones? I've encountered just that frustration in working my way around books and papers whose authors felt that only a small fraction of the equations were splendid enough to deserve a number. Like page numbers, equation numbers help you to hunt down the one you're looking for.

There are many ways to signal that an equation is important without depriving the unimportant equations of the numbers that help you find your way to the important ones. The best way is to write so vividly that it is obvious to the reader that something really noteworthy is about to appear on the page. But even without verbal fanfares for wonderful equations, there are plenty of other devices less disruptive than not numbering the undeserving ones. You can use stars for your prize equations, you can box them, or you can put the equation number in boldface.

3. Another writer argued against the Math Is Prose rule, because it was absurd to insist that *all* equations required punctuation marks, and because almost all punctuation of equations consisted of commas or periods, which served no function other than redundantly directing the reader to pause before continuing past the equation. My reply pointed out that he was

simply wrong when he asserts that virtually the only punctuation required by equations consists of commas and periods. Equally common, and just as important, is the legitimate absence of any punctuation mark, a degenerate form of punctuation that can reveal much about the relation of the equation to the text that follows, but only if periods and commas have also been provided in contexts that require them. Why should the reader have to guess or deduce whether what follows the equation is a new sentence, a new clause, or an extension of the clause that was in progress when the equation made its appearance? Reading mathematical analysis is hard enough without depriving readers of the kinds of clues available to readers of any nonmathematical text.

7

What's wrong with these elements of reality

The subject of Einstein–Podolsky–Rosen correlations—those strong quantum correlations that seem to imply "spooky actions at a distance"—has just been given a new and beautiful twist. Daniel Greenberger, Michael Horne, and Anton Zeilinger have found a clever and powerful extension of the two-particle EPR experiment to *gedanken* decays that produce more than two particles [1]. In the GHZ experiment the spookiness assumes an even more vivid form than it acquired in John Bell's celebrated analysis of the EPR experiment, given over 25 years ago [2]. The argument that follows is my attempt to simplify a refinement of the GHZ argument given by the philosophers Robert Clifton, Michael Redhead, and Jeremy Butterfield [3].

Consider *three* spin-½ particles, named 1, 2, and 3. They have originated in a spin-conserving *gedanken* decay and are now *gedanken* flying apart along three different straight lines in the horizontal plane. (It's not essential for the *gedanken* trajectories to be coplanar, but it makes it easier to describe the rest of the geometry.) I specify the spin state $|\Psi\rangle$ of the three particles in a time-honored manner, giving you a complete set of commuting Hermitian spin-space operators of which $|\Psi\rangle$ is an eigenstate.

Those operators are assembled out of the following pieces (measuring all spins in units of ½ $\hbar$): σ^i_z, the operator for the spin of particle i along its direction of motion; σ^i_x, the spin along the vertical direction; and σ^i_y, the spin along the horizontal direction orthogonal to the trajectory. (Any three orthogonal directions independently chosen for each particle would do. But we're going to be *gedanken*-measuring x and y components of each particle's spin, so it's nice to think of the x and y directions as orthogonal to the direction of motion, since the components can then be straightforwardly measured by passage through a conventional Stern–Gerlach magnet.) The complete set of commuting Hermitian operators consists of

$$\sigma_x^1 \sigma_y^2 \sigma_y^3, \quad \sigma_y^1 \sigma_x^2 \sigma_y^3, \quad \sigma_y^1 \sigma_y^2 \sigma_x^3. \tag{1}$$

Even though the x and y components of a given particle's spin anticommute—a fact of paramount importance in what follows—all three of the operators in (1) do indeed commute with one another, because the product of any two of them differs from the product in the reverse order by an even number of such anticommutations. Because they all commute, the three operators can be provided with simultaneous eigenstates. Since the square of each of the three is unity, the eigenvalues of each are $+1$ or -1, and the 2^3 possible choices are indeed just what we need to span the eight-dimensional space of three spins-½.

For simplicity of exposition let's focus our attention on the symmetric eigenstate in which each of the operators (1) has the eigenvalue $+ 1$. (Its state vector is $|\Psi\rangle = \left(1/\sqrt{2}\right)\left(|1,1,1\rangle - |-1,-1,-1\rangle\right)$, where 1 or -1 specifies spin up or down along the appropriate z axis, but you don't need to know this. I'm only telling you because discussions of EPR always write down an explicit form for the state vector and I wouldn't want you to think you were missing anything.) Because the spin vectors of distinct particles commute component by component, we can simultaneously measure the x component of one particle and the y components of the other two (using three Stern–Gerlach magnets in three remote regions of space). Since the three particles are in an eigenstate of all three operators (1) with eigenvalue unity, the product of the results of the three spin measurements has to be $+1$, regardless of which particle we single out for the x-spin measurement.

This affords an immediate application of the EPR reality criterion [4]: "If, without in any way disturbing a system, we can predict with certainty the value of a physical quantity, then there exists an element of physical reality corresponding to this physical quantity." The "element of physical reality" is that predictable value, and it ought to exist whether or not we actually carry out the procedure necessary for its prediction, since that procedure in no way disturbs the system associated with it.

Because the product of the results of measuring one x component and two y components is unity in the state $|\Psi\rangle$, we can predict with certainty the result of measuring the x component of the spin of any one of the three particles by measuring the y components of the two other, faraway particles. For if both y components turn out to be the same then the x component, when measured, must yield the value $+1$; if the two y components turn out to be different, the subsequently measured x component will necessarily yield the value -1. In the absence of spooky actions at a distance or the metaphysical cunning of a Niels Bohr, the two faraway y-component measurements cannot "disturb" the particle whose x component is subsequently to be measured. The EPR reality criterion therefore asserts the existence of elements of reality m_x^1, m_x^2, and m_x^3, each having the value $+1$ or -1, each waiting to be revealed by the appropriate pair of faraway y-component measurements.

In much the same way, we can also predict the result of measuring the y component of the spin of any particle with certainty, by measuring one x component and one y component of the spins of the other two. There are thus elements of reality m_y^1, m_y^2, and m_y^3, with values $+1$ or -1, also waiting to be revealed by faraway measurements. All six of the elements of reality m_x^i and m_y^i have to be there, because we can predict in advance what any one of the six values will be by measurements made so far away that they cannot disturb the particle that subsequently does indeed display the predicted value.

This conclusion is, of course, highly heretical, because σ_x^i does not commute with σ_y^i —in fact the two *anti*commute—and therefore they cannot have simultaneous values. (The operators (1) are nicely chosen to hide this failure to commute, since the anticommutations always occur in pairs.) But heresy or not, since the result of either measurement can be predicted with probability 1 from the results of other measurements made arbitrarily far away, an open-minded person might be sorely tempted to renounce quantum theology in favor of an interpretation less hostile to the elements of reality.

In the GHZ experiment, however, as in Bell's version of the EPR, the elements of reality are demolished by the straightforward quantum mechanical predictions for some additional experiments, entirely unencumbered by accompanying metaphysical baggage.

In the GHZ case the demolition is spectacularly more efficient. Suppose, heretically, that the elements of reality really do exist in each run of the experiment. While we cannot know all six of their values, those values are constrained by the fact that the values of $\sigma_x^1 \sigma_y^2 \sigma_y^3$, $\sigma_y^1 \sigma_x^2 \sigma_y^3$, and $\sigma_y^1 \sigma_y^2 \sigma_x^3$, all unity in the state $|\Psi\rangle$, are given by the values of the corresponding products $m_x^1 m_y^2 m_y^3$, $m_y^1 m_x^2 m_y^3$, and $m_y^1 m_y^2 m_x^3$. But if these latter three quantities are unity, so is their combined product. Since each individual m_y^i is either +1 or −1 and each occurs twice in the combined product, that combined product is just $m_x^1 m_x^2 m_x^3$. So the existence of the elements of reality implies that should we choose to measure the x components of all three spins in the state $|\Psi\rangle$, the product of the three resulting values must once again be +1.

The value of that product can also be determined without invoking disreputable elements of reality by a simple quantum mechanical calculation, since it is just the result of measuring the Hermitian operator

$$\sigma_x^1 \sigma_x^2 \sigma_x^3. \tag{2}$$

You can easily check that this operator also commutes with all of the operators (1): Once again the number of anticommutations is always even. This is encouraging, for if the value of the operator (2) in the state $|\Psi\rangle$ is invariably to be +1, it had better also have $|\Psi\rangle$ for an eigenstate, a requirement that is guaranteed by its commuting with all three members of the complete set of commuting operators (1) whose eigenvalues define $|\Psi\rangle$.

However:

Not only does (2) commute with each of the operators (1), but you can easily check that it is a simple explicit function of them, namely, *minus* the product of all three. The (crucial) minus sign arises because here, at last, in bringing the pairs of operators σ^i_y together to produce unity, one runs up against an *odd* number of anticommutations of σ^i_ys with σ^i_xs. Since $|\Psi\rangle$ is an eigenstate with eigenvalue +1 of each of the operators (1), it is therefore indeed an eigenstate of the operator (2), but with the *wrong* eigenvalue, opposite in sign to the one required by the existence of the elements of reality.

So farewell elements of reality! And farewell in a hurry. The compelling hypothesis that they exist can be refuted by a *single* measurement of the three x components: The elements of reality require the product of the three outcomes *invariably* to be +1; but *invariably* the product of the three outcomes is −1.

This is an altogether more powerful refutation of the existence of elements of reality than the one provided by Bell's theorem for the two-particle EPR experiment. Bell showed that the elements of reality inferred from one group of measurements are incompatible with the *statistics* produced by a second group of measurements. Such a refutation cannot be accomplished in a single run, but is built up with increasing confidence as the number of runs increases. Thus in one simple version of the two-particle EPR experiment[1] the hypothesis of elements of reality requires a class

1 I described this version in *Physics Today*, April 1985, page 38.

of outcomes to occur at least 55.5% of the time, while quantum mechanics allows them to occur only 50% of the time. In the GHZ experiment, on the other hand, the elements of reality require a class of outcomes to occur *all* of the time, while quantum mechanics *never* allows them to occur.

It is also appealing to see the failure of the EPR reality criterion emerge quite directly from the one crucial difference between the elements of reality (which, being ordinary numbers, necessarily commute) and the corresponding quantum mechanical observables (which sometimes anticommute).

I was surprised to learn of this always-vs-never refutation of Einstein, Podolsky, and Rosen. After all, quantum magic generally flows from the fact that it is the amplitudes that combine like probabilities rather than the probabilities themselves. But when the probabilities are zero, so are the amplitudes. Guided by such woolly thinking, and the failure of anybody to strengthen Bell's result in this direction in the ensuing 25 years, I recently declared in print [5] that no set of experiments, real or *gedanken*, was known that could produce such an all-or-nothing demolition of the elements of reality. With a bow of admiration to Greenberger, Horne, and Zeilinger, I hereby recant.

References

1. D, M. Greenberger, M. Horne, A. Zeilinger, in *Bell's Theorem. Quantum Theory, and Conceptions of the Universe*, M. Kafatos (ed.), Dordrecht, The Netherlands: Kluwer (1989), p. 69.
2. J. S. Bell, *Physics* 1, 195 (1964).
3. R. K. Clifton, M. L. G. Redhead, J. M. Butterfield, "Generalization of the Greenberger-Horne-Zeilinger Algebraic Proof of Nonlocality," submitted to *Found. Phys.*
4. A. Einstein, B. Podolsky, N. Rosen, *Phys. Rev.* 47, 777 (1935).
5. N, D. Mermin, in *Philosophical Consequences of Quantum Theory*, J. T. Crushing, E. McMullin (eds.), Notre Dame University Press (1989), p. 48.

7. Postscript

1. *Elements of Reality* is the first of a few columns[2] written for a reader with an elementary knowledge of quantum mechanics. It contains only two displayed and numbered equations, but is more technical than most articles in *Physics Today*. Google Scholar, (as of early 2015) lists over three hundred citations of *Elements of Reality* in the scientific literature, making it my thirteenth-most-cited technical paper.

2. If I had written *Elements* in a less exuberant style I could have published it in a major technical journal like *Physical Review Letters*, but only in *Physics Today* could I freely express my enthusiasm for the beauty of the argument. My opinion was shared by John S. Bell, whose famous 1964 paper is the first of all such arguments. I sent him a copy of the column before it appeared in print. He replied "I am full of admiration for your three-spin trick."

3. Today when people talk about Greenberger, Horne, and Zeilinger (usually abbreviated to GHZ[3]) they almost always have in mind the simplification of their argument that I published as a *Reference Frame* column. Without the large readership of *Physics Today* and the informal, readable style of *Reference Frame*, GHZ might not have been noticed.[4] Danny Greenberger once introduced me to a friend as "the man who made me famous."

---

2 The others are Chapters 16, 23, 28, and 29.
3 GHZ should not be confused with GHz, the abbreviation for gigahertz.
4 Recall how nobody noticed several years of "Lagrangean" (Chapter 1).

8

What's wrong with these reviews

Se vuol ballare, signor Contino, il chitarino, le suoneró.

– Figaro

The story you are about to read is true. The names, to be sure, have been changed, but not to protect the innocent. Professor Mozart would have been only too pleased for me to use his real name, but we agreed that you, dear reader, might think you were reading a piece of special pleading on behalf of a particular person. Far from it. The point of my tale is not that one physicist has been badly dealt with by the National Science Foundation; indeed Mozart has been very well treated over the years and considers himself lucky to have escaped this time with the wherewithal to keep supporting a solitary graduate student. No, the point of this tale is to illustrate more vividly than reams of surveys or statistics could possibly convey what has happened to NSF support for research in condensed-matter theory.

Condensed-matter theorists have been maintaining for some time, in these pages and even on the op-ed page of *The New York Times*, that their discipline is being starved by NSF. In reply, NSF has insisted that things are hard all over, and scientists from all over have tended to agree. So I offer the tale of Professor Mozart as a benchmark against which to test the plight of your own field. Have things like this been happening to people in your corner of science?

My tale begins several months ago, when NSF phoned Professor Mozart to tell him that a small condensed-matter theory grant he shared with Professor Beethoven would be renewed, but with a 20% cut—Mozart was to lose 30% and Beethoven 10%.

Mozart was told that four of the five reviewers had given the proposal E's (the highest possible rating) and one a G (two notches down from the top or two notches up from the bottom, depending on the case you're trying to make). Mozart was informed that he, not Beethoven, was responsible for this blemish, and was urged to get his act together if he expected to get any support at all in the next round.

Mozart, who knows perfectly well how things stand in condensed-matter theory and had been expecting far worse, was actually relieved by this turn of events. He was downright pleased to have been presented with an irrefutable piece of evidence that the point had been reached where a set of ratings just one reviewer short of perfection could lead to a 30% budget cut in a program that was modest to begin with, and he entertained several nearby colleagues with this latest horror story. It traveled quickly around and soon showed up in Bob Park's APS computer newsletter, *What's New*, as an anecdote about a man who had won an APS prize for "outstanding contributions to physics" and then had his NSF grant cut by 30% after getting four E's out of five on his proposal.

Shortly after his anonymous appearance in *What's New* Mozart received a call from NSF. Somebody had gone to the archives, looked up recent citations and tracked him down. How could he have spread such misleading information? There was already too much hysteria in the air, and this kind of irresponsible talk only fanned the flames. When Mozart saw the actual reviews he would realize how fortunate he was to have been renewed at all! Distinctly chastened, Mozart said that when he received the reviews he would insist that Park run a correction if that seemed called for.

So when the reviews arrived Mozart opened them with grim foreboding, prepared to see in black and white the unvoiced doubts that sometimes tormented him in private midnight moments. What he first noticed was that *What's New* had indeed misrepresented the bare facts of the case. The proposal was, to be sure, given four E's and a G, but one reviewer addressed only Mozart and one only Beethoven. Mozart's actual grades were only three E's and a G. Beethoven (who was still cut 10%) got four E's. (Beethoven was furious, but that's another story.)

Equally alarming, Mozart had been told in the admonitory phone call that one of the three E reviews that addressed both Mozart and Beethoven should be dismissed as superficial, and there it was, at the top of the pile: a four-liner saying only that Mozart and Beethoven were both well-known condensed-matter theorists whose earlier work for NSF was "of the highest quality and covers a wide range of topics." The proposed research was "interesting, deserves support, and the reviewer has no doubt that significant contributions to the field will be forthcoming as it is carried out." Mozart says he couldn't agree more that this is the most flagrant kind of two-person E-boilerplate. We trust that NSF has expunged the author from its roster of peers.

So there was poor Mozart, down to two acceptable E's. The first nonsuperficial E review began with a reference to Mozart and Beethoven's joint "long track record of doing excellent research, including the recent work" on the previous grant. But the next sentence sent a chill through Mozart: "Mozart's proposal for future work appears feasible, but is a bit sketchy in parts, and not overly ambitious." So the E was a slip of the pen, or perhaps a manifestation of Beethoven's formidable coattail powers? No, not at all! "This is not troubling," continued the reviewer, "because I am confident based on his superlative record that he will make important contributions to our understanding of [*deleted to preserve anonymity*], although at this point I do not know what they will be (and apparently neither does he in great detail)."

"What do you make of that?" I asked Mozart.

"The reviewer is absolutely correct," he said. "I never know what I'm going to do until I've done it; otherwise it wouldn't be research."

"Mozart," the reviewer concluded, "thinks very clearly and elegantly. I strongly recommend funding his research on the basis that the NSF should be funding some very basic research and that he is one of the best and most productive scientists in the country in his field."

His spirits more than a little buoyed, Mozart moved on with diminished but still significant trepidation to the second nonsuperficial E review, which said of Mozart and Beethoven jointly: "I give this proposal the highest rating. It is difficult for me to imagine its not being funded." The reviewer then considered Mozart on his own: "The work by Mozart on [deleted] was truly important to the field of [deleted]. It established a general framework for [deleted] that has become the standard. The discovery that [deleted] was an elegant and surprising result of this general investigation of [deleted]. The studies of [deleted] greatly enhanced our understanding of these unusual materials. The review article with Boccherini on [deleted] was a service to the physics community." And then, returning to Mozart after waxing equally rhapsodic about the work of Beethoven: "Mozart's continued research on [deleted] addresses interesting and unanswered questions in these fields."

You can imagine Mozart's delight in reading all this, after his telephonic reprimand from NSF, But what about that smoking G? Here it is:

"Mozart's ideas are reasonable and worthwhile, but I don't see anything exciting or new. He proposes further work on [deleted] which will fill out our understanding of these materials. If the NSF budget for condensed-matter theory were larger, I would recommend funding, as Mozart is a productive, highly competent physicist. However, if you have proposals from young scientists with interesting new ideas that are put forth in considerable detail (i.e., longer than the two pages Mozart bothers to write here), I think it is clear where your priorities should be. For a fundable proposal nowadays I expect to see 10–15 pages of well-thought-out ideas, some of which involve finishing up old work, but the bulk of which involve new directions. I don't see that here."

"What's this?" I asked Mozart. "You wrote a two-page proposal?"

"Well," he replied, "the expository text for my half of the proposal, excluding the general preamble, lists of publications and that sort of thing, ran to about ten pages. As the reviewer who mentioned 'sketchiness' seemed to understand, I've never been good at anticipating what I'll be up to next, so I wrote the 'Work Proposed' section in the form of a brief commentary on the 'Work Accomplished' part. It seemed efficient to lay the groundwork for my best guess at where I would be heading in the next three years in the context of where I had already been. Had I realized what I was letting myself in for, I could easily have redistributed a considerable amount of text between the two sections, and the next time around I certainly will. But the criticism that I did not spell out all the discoveries I intended to make in the next three years is absolutely correct.

"Anyway," Mozart concluded amiably, "I entirely agree that in these hard times the interesting new ideas of young scientists should have the highest priority. It's just unfortunate that we have to sacrifice the reasonable and worthwhile ideas of productive and highly competent older people to do it."

Mozart asked me to emphasize that the NSF program officers in the condensed-matter theory section are doing a heroic job, having to make impossibly fine distinctions to distribute grossly inadequate resources in a rapidly expanding and exceptionally productive field of science. I would only add that it is sad that the funding crisis in condensed-matter theory has now reached the point where reviews like those I've just quoted (the reading of which filled Mozart with pride and pleasure, though he had been led to dread their arrival) can form the basis for a grim warning from a program officer that their recipient is lucky to be renewed with a mere 30% cut. What makes this even sadder is that it still seems to be the party line at NSF that no special funding crisis exists in condensed-matter theory. Equally distressing is the message that you had better spell out in lengthy and explicit detail where your research is heading over the next three years, preferably jumping off in an entirely new direction, if you wish the people at NSF to continue to support you. I hope they haven't really adopted this recipe for opportunism and mediocrity, but it worries me that they may now be so desperate to find any excuse to turn down proposals in condensed-matter theory that they will start believing their own rationalizations.

8. Postscript

1. Since World War II the federal government has supported science at universities in the United States. Originally this generosity was motivated by national defense, and the unquestionable military relevance of, for example, the huge federal expenditure on nuclear physics at the Manhattan project. In subsequent years it became clear that maintaining American leadership in fundamental science had important economic benefits for the nation, and that, with some outstanding exceptions, private industry was not well suited to support so broad a goal, focused, as it necessarily was, on short-term marketable applications. By the early 1990s industrial support for fundamental science was rapidly vanishing, and the pressure on federal agencies was becoming acute.

2. Since I arrived at Cornell in 1964 I had shared a modest grant from the U. S. National Science Foundation with a small and varying number of Cornell colleagues in the field of theoretical condensed-matter physics.[1] Early in 1990 two of us applied to renew our grant, and my half of the proposal received a review which struck me as paradigmatic of the difficult new conditions. The text of the review provided a basis for a discussion in *Reference Frame* of the current crisis in science funding. But there was a problem. If I identified it as a review of my own proposal,

---

1 The term "condensed-matter physics" was introduced in the 1960s as a replacement for "solid-state physics," on grounds of accuracy: solid-state physicists were also interested in liquids and other nonsolid forms of bulk matter. I thought at the time and continue to think this was a damaging move for the field. Music lovers around the world had learned that solid-state devices were wonderful. But "condensed matter" suggested condensed books, condensed milk, and other disagreeable things.

that would be unprofessional and self-interested. Indeed, *Physics Today* would almost certainly have refused to publish it. So I invented what was obviously a pseudonym—Professor Mozart—for the recipient of the reviews, and that solved the problem.

3. When I wrote *Reviews* I thought Mozart was putting in a one-time appearance. But in subsequent columns he expressed outrageous views on several other issues that I was not brave enough to offer as my own opinions. As a hint that I was publishing excerpts from the official response of NSF to my own research proposal, I took as an epigraph Figaro's response to the misbehavior of his boss, Count Almaviva: "If you want to dance, my lord, I'll play the guitar for you."

4. Surprisingly, I received just one critical response to my column. It came from a former Cornell physics graduate student, who viewed it as self-serving both individually (why does Professor Mozart think anybody owes him financial support?) and collectively (why is condensed-matter physics singled out for special pleading?). Several friendlier letter writers pointed out that things were just as bad, if not worse, in their own fields.

5. Shortly after this column appeared, my friend and Cornell colleague Neil Ashcroft, with whom I had written a well-known solid-state physics text, reported with surprise and amusement that when he visited other universities to give a lecture, included in his introduction to the audience was often the information that he was, in fact, Professor Mozart. A strange but widespread jump to the wrong conclusion. He was not even Professor Beethoven. This was one of the first indications I had of just how widely read *Reference Frame* was among physicists.

9

What's wrong with those epochs

Ed hai corragio di traitor scherzando un negozio si serio? [And you have the nerve to joke about so serious a business?]

– Susanna

M y amiable friend Professor Mozart dropped by the other day. Now that his NSF grant has been cut way back, he has more time to think about things, and it's a pleasure to chat with him. Some of his views, though, are more than a little peculiar, as the following conversation clearly reveals.

"I have to admit," Mozart began sadly, "that particle physics over the last 40 or 50 years has been a disappointment. Who would have expected that in half a century we wouldn't learn anything really profound?"

"Nothing profound?!" I exploded. "What about parity nonconservation? What about the breakdown of time-reversal symmetry?"

"To be sure," sighed Mozart, "we've learned that left can be distinguished from right and that time past is different from time future. But most ordinary people knew the difference between left and right all along, and who except the most highly trained physicists—temporarily, it now turns out—ever doubted for a moment that they could tell the future from the past? So establishing that the asymmetry is really there after all is certainly commendable. But about really serious problems we've discovered nothing—nothing whatsoever about the central puzzle."

"And just what might that puzzle be?" I urged, for he seemed in danger of succumbing to an attack of melancholia.

He revived. "All particle physics has taught us about the central mystery is that quantum mechanics still works. Perfectly, as far as anybody can tell. What a letdown!"

"Letdown? It's a triumph!"

"Letdown!" he insisted. "Think of the previous half-century, when we went down from the macroscopic by seven or eight orders of magnitude. What delicious confusion! All the verities of the preceding two centuries, held by physicists and ordinary people alike, simply fell apart—collapsed. We had to start all over again, and we came up with something that worked just beautifully but was so strange that nobody had any idea what it meant except Bohr, and practically nobody could understand *him*. So naturally we kept probing further, getting to smaller and smaller length scales, waiting for the next revolution to shed some light on the meaning of the old one. But what happened? For 65 years, since 1925, we've been probing, at finer and finer levels. That's more than a quarter of the time between 1685 and 1925. And more of us have been working on the problem than the world's entire supply of physicists between Newton and Bohr. As for our *funding* [poor old Mozart still can't keep his mind off funding for very long], well our funding has absolutely dwarfed all the combined funding from Bohr clear back to Archimedes.

"But what have we to show for it? We got from atoms down to the nucleus, and quantum mechanics still worked perfectly. Inside the nucleus it still worked perfectly. Inside the nucleon it's still working perfectly. Here we are today, another seven or eight orders of magnitude down beneath the level of the old revolution, and nothing fundamentally new is in sight—to be sure, some lovely new Lagrangians, but not the slightest trace of a hint of anything better than quantum mechanics. Disappointed? You bet." He picked up my old copy of Bjorken and Drell and thumbed morosely through it.

"But look what else we've learned in the process," I protested. "There's the connection between particle physics and cosmology, that astonishing link between the biggest and smallest of things.

We're now studying the very earliest moments after the Big Bang! Even if we haven't managed to shed light on the great issues of principle that preoccupy you, surely we're learning a lot about the raw facts of nature. Why, we can recreate in the laboratory the earliest seconds—the earliest milliseconds—perhaps even the earliest microseconds, back when the whole universe wasn't much bigger than the solar system. Doesn't that make you proud?"

"No," Mozart smiled wanly, "not me. Just calm down, and ask a few old questions. For example, what is time? That's easy: Time is what clocks tell. And what are clocks? Objects you can find in the environment or make out of things you find that behave in a periodic way so you can count cycles. And what was the environment like in that first microsecond or two? Hot, I'll tell you! Spectacularly hot. So hot that the characteristic frequencies of anything worthy of the name "clock" were just unbelievably high. So high that for those clocks a microsecond was just eons and eons of time—probably as long a time for those clocks as the age of the universe is for us today. And that's hardly surprising, since, after all, a microsecond *was* the age of the universe, way back then.

"The fact is," he continued briskly, reverting to his more familiar professorial manner, "that a linear time scale makes no sense in cosmology. It gets us all excited about getting back to the beginning when we're really nowhere near it and never will be able to get anywhere near it. We can only get there in constant seconds, but it's current seconds that matter—the seconds ticked off by the feasible clocks of the current epoch. So all those constant milliseconds back then contained vast ages of current seconds, within which events crept in their petty pace from femtosecond to femtosecond... ." He subsided back into gloom.

"OK, W. A., so the time scale should be expanded. But why does that matter?"

Mozart gave me the reproachful look he reserves for students who aren't really trying. "If we say we're chasing the behavior of matter down to the earliest milli-, micro-, or nanosecond, then we think we're getting somewhere—revealing the great essence of things at the very earliest moments. But I say all we're doing is getting glimpses of epoch 3, epoch 4, and epoch 5, each with its own characteristic phenomenology, each more fleetingly revealed, with literally countless ranks of prior epochs waiting to tease us with still more faintly discernible fragments of their characteristic features. We are, my friend, striving after ever more crude glimpses into the phenomenology of the ever more remote past. Particle physics has become the archeology of physics. Every time we go up a few orders of magnitude in energy we're able to start constructing the phenomenology of a still earlier epoch. To be sure, that gives us more insight into the epoch that followed it. But beneath the last layer we have learned a little about, there will always be another about which we know nothing.

"Not that the enterprise is without great merit. Somebody has to dig up the pottery shards, note what layers they come from, and try to make intelligent inferences about what they tell us of the era that produced them. Still, it's tame stuff compared with"—here he brightened perceptibly—"the broad and sophisticated views ordinary physics is giving us of the intricate phenomenology of the living present. *De gustibus non est disputandum...*." And a smile of admiration for the wonders of the present epoch brightened his face.

I was glad to see him recover his customary good spirits, but his smugness irritated me. "Hold on, W. A.—when you think about your beloved present epoch you can't avoid the great lesson particle physics has taught us: that everything—absolutely everything—the Sun, Mount Katahdin, you, me, barium titanate, mesoscopic heterostructures—we're all made out of quarks and leptons. That's all there is. Just quarks and leptons, put together in different ways. So what can be more fundamental than learning more about them? The answer to any question you can ask goes back to quarks and leptons."

"That," shot back Mozart, roused from his reverie, "is like wondering what makes Shakespeare so powerful. One day it hits you that everything he ever wrote is made up of words. So you start looking at the plays as bunches of words and make some interesting discoveries. There's only a finite number of these building blocks—less than 50 000. You can order them by frequency, or by the frequency of consecutive pairs, and you discover that Shakespeare has his own characteristic frequencies, which are different from those of other writers, who have their own patterns, and you can even write computer programs that take a text and tell you whether or not it's by Shakespeare. So you think you're getting somewhere, toward a sense of what makes Shakespeare special.

"But then somebody else comes along with another discovery: The words are all made out of letters, and there are only 26 of those—maybe as many as 100 if you want to include punctuation and capitalization. So we've enormously reduced the number of fundamental units out of which Shakespeare's plays are composed. Of course the letter frequencies aren't as useful as the word frequencies in distinguishing Shakespeare from the New York penal code, but they do help in telling him from Dante, and anyway, the words and word–word correlation functions that were so promising a line of attack can all be expressed as higher-order multiletter correlations, so all the information is still there in the letters. Since they're the basic constituents of the words, they have to be more fundamental, more important to study, more exciting a way to approach Shakespeare.

"And then somebody notices something very important about these letters—that they're made up of very similar lines. For example, if you take two parallel vertical lines and connect them with a horizontal line you get an 'H,' but if the line is diagonal, you get an 'N,' so it's all just the arrangement of an even smaller number of little lines. But somebody else discovers ASCII coding and realizes that all of Shakespeare is built up out of just two units: 0 and 1. Then there are the phenomenologists, who say no, that's not the point—it's really two fundamental substances, paper and ink, and the key to Shakespeare lies in the way the ink penetrates the paper."

Mozart sighed deeply. "There are few facts less interesting than the fact that everything is made out of quarks and leptons, even if it does survive the next round of excavations. No, what's important about particle physics is the wonderful archeology, for its own sake. It's admirable that while most of us are preoccupied with puzzling out and admiring the extraordinary intricacies organized structures of the present epoch present us with, many dedicated souls remain committed to digging out the shards and fragments of the earlier epochs. The time will surely come, at several of the more shallow levels, when they will succeed in assembling their shards into entire beautiful pots." Here he smiled the smile of one who deals in bone china, Wedgwood bowls, Tiffany lampshades, and crystal menageries. "And it is my hope," he added benignly as he sailed off toward the elevator, "that someday they will, after all, discover something genuinely profound. Something that teaches us a little more about the serious problem."

"Wait," I shouted as the elevator doors closed. "What about the electroweak unification?"

9. Postscript

1. I intended *Epochs* to be a meditation upon the remarkable fact that quantum mechanics, although developed to account for the behavior of matter at the atomic scale, continues to work perfectly when applied inside the atomic nucleus, even though the diameter of a nucleus is some hundred thousand times smaller than the diameter of an atom. Quantum mechanics even provides the language necessary to describe the quarks and gluons that make up the deep internal structure of each proton and neutron.

Why should a radical new point of view, developed entirely from evidence arising on the atomic scale, remain valid when applied to phenomena taking place on vastly tinier length scales? Most physicists seem unsurprised at this. While it makes sense to apply quantum mechanics to smaller and smaller things, waiting to see where and how it breaks down, few physicists seem to have entertained the possibility that it actually might break down. Its spectacular success at length scales many millions of times smaller than those for which it was designed seems to have been taken for granted.

Being reluctant to chide my colleagues for their complacency, it occurred to me that this was another job for Professor Mozart. I could be the physicist-on-the-street, enchanted by the spectacular achievements of particle physics, and Professor Mozart would reprimand me for failing to notice that on the really deep questions—the meaning of quantum mechanics and whether it would break down on shorter length scales—we had learned nothing whatever beyond the (admittedly important) fact that quantum mechanics was working far better than we had any right to expect.

Given the podium, Professor Mozart, to my surprise, took advantage of it to press his private agenda: how unappreciated his own field of condensed-matter physics was in the broader physics community, and how over-hyped were some of the much ballyhooed achievements of particle physics. My conversation with Mozart wanders into some good-natured (or so I thought) teasing of particle physicists for their naively reductionist claim to be *the* most fundamental field of science.

2. Realizing that I might be forming a habit of beginning my column with a quote from a Mozart opera, *Physics Today* asked me to supply a translation this time. Having always regarded epigraphs from languages I didn't speak as entertaining puzzles, I resisted. They insisted. I'm proud of having translated *corragio* as "nerve," rather than "courage."

3. It turned out that it did indeed take nerve to joke about so serious a business. Some particle physicists were not amused. To my astonishment, my good-natured teasing of my particle-physicist colleagues was taken by some to be an attack on the Superconducting Super Collider.

The SSC was authorized by Ronald Reagan in 1987 and approved by Congress in 1988. It was to be the most powerful particle accelerator ever built, and indeed, it would have been substantially more powerful than the current world champion, the Large Hadron Collider (LHC) that in 2010 finally began operating at CERN, outside of Geneva, and in 2012 announced to an admiring world the discovery of the Higgs boson.

After an intense national competition, a site for the SSC was selected in Waxahachie, Texas. During construction the projected cost kept growing, and when it exceeded three times the original estimate, after billions of dollars had already been spent and thousands of people had moved to Waxahachie, in the middle of nowhere, the project was cancelled by Congress in 1993. Today the site is a colossal abandoned ruin, epitomizing for many the end of the half century of American dominance in world science that began with Hitler firing most of Germany's finest scientists, and concluded with the retreat of Congress from the pursuit of scientific excellence for its own sake.

The project was controversial among physicists. Those who were not in particle physics suspected that its financial demands were at least partly responsible for the increasingly tight constraints on federal support for their own projects. And there was resentment at the overblown claims some particle physicists were making on behalf of particle and accelerator physics, appropriating achievements of quite unrelated branches of physics as spin-offs of their own work.

4. It was at this time of crisis, in 1991, that my friendly joshing of some of the sillier pretensions of my particle-physics friends appeared, only three months after my complaint about funding difficulties in my own field of physics (Chapter 8). In February 1991, *Physics Today* published two letters critical of *Epochs*. I learned that the device of putting my more controversial views into the mouth of a fictional character liberated my critics as well as myself. One letter said that Professor Mozart "seemed

to have lost most of his marbles … If he put drivel about 'the central mystery' into his NSF proposal, it's no wonder that his grant was cut." Reference was made to the "blot on the landscape of intellectual history that will disgrace all of basic science" if the SSC were not completed, a disaster that Professor Mozart surely would not wish to encourage, once he "recovers his wits." Another letter, noting that whom the gods would destroy they first make mad, suggested that Professor Mozart was "suffering from some ailment."

I give below those parts of my reply to these onslaughts that bear on the development of the character of Professor Mozart and his remarks in subsequent chapters:

Professor Mozart loves quarks and can hardly wait for particle physicists to dig down to how the world works at subquark distances. But confinement doesn't befuddle him the way quantum mechanics does. "*Conceptually* it's no stranger than Hooke's law," he says, "except for the absence of a cutoff."

Professor Mozart should not be held responsible for my own exuberant merging of the goals of cosmology and particle physics. Mozart's point was more subtle: that the possibility of an endless hierarchy of shorter and shorter time scales in the early universe, each with its own characteristic features, suggests the analogous possibility of a hierarchy of shorter and shorter length scales, each with a newer and more "fundamental" particle phenomenology than the one above it. Thus he finds in cosmology some serious warnings about the path particle physicists are pursuing.

And who said anything about the SSC? Can no opportunity be lost to praise its scientific, intellectual, and morally uplifting qualities? As it happens Professor Mozart is a big supporter of the machine. He's filled with curiosity about what will turn up in the next layer and delighted that the public is willing to invest billions in a purely intellectual exercise, with major spin-offs for cosmology. "Certainly it would be tragic to stop digging now," he insists, "comparable to the loss of the Great Library at Alexandria. Whatever the layer at which we finally have to stop, it will be tragic."

Nor is Mozart worried about the drain the project might impose on the rest of science. He says his productivity has actually increased since his funding was cut. "Fewer reports to file, fewer graduate students to worry about, and more time to follow my nose, wherever it takes me." He is, of course, a theorist, but as for the experimentalists, "A temporary return to string and sealing wax on the kitchen table would refresh them all."

5. A further reference to *Epochs* can be found in a letter of mine to the editor of *Physics Today* (February 1992). At this stage in the sad story of the SSC, desperate efforts were being made to persuade other countries to share the ever-growing cost of the project:

> I just got an email message from Kazan containing, among other things, a request that I forward the following letter to the editor of *Physics Today*:
>
>> In a recent book review (June 1991, page 108) my good friend Philip W. Anderson says, "It is even possible for David Mermin to *complain* about the universe being boring quantum mechanics all the way down." While Mermin does complain a lot, in this case he actually said that the success of quantum mechanics all the way down to where we've got is "a triumph." It was I who expressed disappointment at this state of affairs. Neither of us finds quantum mechanics boring.
>>
>> May I also take this opportunity to say as emphatically as I can that I am not now nor have I ever been a pseudonym for Neil W. Ashcroft.
>>
>> William A. Mozart
>
> The author of the letter has been abroad for some time now, trying to raise funds for the SSC. He says the Tatars are wild for Waxahachie but a little short of hard currency. He hopes, however, to persuade them to contribute rugs.

10

Publishing in Computopia

I have just finished writing a short technical article that ties together two old, important, and previously unrelated results in a surprising way that simplifies and elucidates them both. It is self-contained and readable, and the formal analysis it employs is extremely simple; it will be cited in textbooks. I make a list of people I think might be interested. The field is a small one, so I start from memory. Next, I go through some conference proceedings I happen to have at hand, to get addresses and catch people I might have missed. I end up with about 50 names and addresses. I fiddle around with the formatting parameters to squeeze the paper into only eight pages, so that my secretary can print it reduced on just two sheets of paper to save postage and copying costs—my grant has been cut. When all the copies are in the mail I return to my computer, unsqueeze the paper, move the footnotes from the bottoms of the pages, where they are easy to read, to the end of the manuscript, as the rules require, and make four more copies that I send off to Physical Review Letters.

What's wrong with this story? What strange, irrational, one might even say unprofessional act have I just described?

Was it wasteful of me to inflict this burden on so many in-baskets, knowing that considerably fewer than half the recipients will look at my paper? Not at all! I will be content if a dozen people take a serious look—that will be enough for my message to propagate—and I have followed the best strategy to bring that about.

Was I, then, foolish to waste paper, postage, and secretarial time in this old-fashioned effort at communication, when email would have done the trick effortlessly? Not yet! Not all my correspondents communicate in that way, I lack the electronic addresses of many of those who do, and surely somebody (though I have no idea who) is paying for all those email transmissions, so the monetary savings to society as a whole may be at least in part illusory.

Was it, perhaps, absurd of me to lavish extreme care on these preprints, rearranging the footnotes for easier reading, and even worrying about the proper choice of fonts to convey to my readers as clearly as possible the relations between the tiny sections and subsections? Not a bit, if this makes it easier for a dozen of them actually to read it.

No, the only peculiar step in the entire process—the only one that casts doubt on my judgment, seriousness of purpose, and moral integrity—was the last one. Why on earth, having done everything necessary, did I then produce four more copies to be sent to *Physical Review Letters*?

Was it for the benefit of people unknown to me, not on my list? Superficially plausible. If the paper goes to the right referees it will be accepted and promptly appear. But there is a significant chance that three months will elapse, after which it will be rejected. It would then be my duty to the people not on my list to resubmit, with a letter explaining why the referees have missed my point. Failing to enlighten them, after four and a half months I would have to struggle on behalf of those unknown to me, not on my list, by requesting new referees. More likely than not, after six or seven months the paper would be accepted. But perhaps not. Half a year having passed, I would then repeat the process with a journal of lower standards and more sympathetic reviewers, and with any luck the paper would be in print not more than a year from my first submission.

During this long interval two or three of the dozen or so who looked at what I mailed them will have mentioned it to others. A few papers will have been circulated commenting on mine, bringing it to the attention of people on different distribution lists. I will get a few requests for copies and will gladly provide them. I and perhaps some of my readers will give talks about it. In short, anybody with more than the faintest interest in the subject will be afforded many opportunities to learn what I have done well before publication.

Well, if not for the benefit of those not on my list, then why indeed submit to *PRL*? For my own benefit, of course! I have a grant proposal pending on the subject of my article. These days it is the first duty of all program officers to reject as many proposals as they can, and a powerful argument against funding mine is that everything of interest to say on the subject has already been said. Having concrete evidence that *PRL* thought otherwise will be valuable in the resubmission.

Is this, then, an efficient way to marshal evidence in support of grant proposals, appointments, promotions, or fellowships? Of course not—it is madness. The time is overdue to abolish journals and reorganize the way we do business.

[Honesty compels me to note that since the above was written *PRL* accepted my paper on the first round and published it promptly, and I received one thank-you note from an enthusiastic reader not on my list. This has no bearing on the points made below. One does not formulate policy on the basis of singular events.]

In a rational world, paper, printing, postage, and *PRL* would never have crossed my mind. I would simply have emailed my essay to a central clearinghouse for posting on its electronic bulletin board. Readers around the world would peruse this bulletin board, which would be organized by the usual categories that document the fragmentation of human knowledge. After calling up for inspection the abstracts of titles that caught their interest, readers could call up copies of entire articles, even printing them out on the rare occasions they promised to repay careful study.

Why don't we do this? Well, not everybody has access to the networks, so it's unfair.

Nonsense! Even fewer people have access to preprints, the current avenue of serious communication; nothing could be more unfair than the way we now do business. By the time the information has diffused to those not on original lists—brief as it may be compared with the time to publication—those on the lists have gone galloping off to the next stage, leaving the unlisted only the less exhilarating pleasures of cleaning up the mess they've left behind. In fact the numbers of the electronically un-plugged-in are diminishing daily, but to solve the problem of universal and immediate access libraries can, for a tiny fraction of the enormous sums they now waste on journals, set themselves up with terminals for perusing the bulletin board and with printers for producing hard copies at no more expense to users than is currently spent on copying machines. The ridiculous five-step process by which my ideas now end up in the pile on your desk (computer file → printer →·keyboard operator → computer file → journal page → copying machine) will collapse to one.

But the papers on the bulletin board would be unrefereed, imposing a new burden on readers.

This is no disadvantage. It would lift an even greater burden from the community of referees, which hardly differs from the community of readers. Truly conscientious referees now have to spend most of their waking hours reading the papers of others, if they are saintly enough to give the job the attention it requires. What would be lost by the disappearance of refereeing? Readers would have to decide for themselves whether a paper was rubbish; that is, they would have to referee (for themselves) only those papers they might really be interested in looking at. Because none of us are saints, current refereeing is so ineffectual that this is already necessary. Under the new system readers would simply have to sift through a somewhat larger pile. Not having to struggle, as an official referee, with many papers one is at best peripherally interested in would more than compensate for this.

What about the validation that publication in a peer-reviewed journal stamps on a paper for those who dispense grants and promotions?

Most papers, of course, now have multiple authors, and the particular contribution of any single one is shrouded in obscurity.

Nevertheless, authors may legitimately require an official assessment of their product for private purposes, so the professional societies will have to maintain panels of reviewers, just as their journals now maintain panels of referees. Those needing such validation (and no others) would submit their manuscripts to the appropriate panel for a grade: A+, A, A–, ..., D–, F. This process of evaluation would be decoupled from any posting of the manuscript on the electronic bulletin board—it would be up to the author to decide whether to post before receiving the grade. There could be one exchange to allow an author to improve the paper for a better grade or to present a case for raising a poor grade. After that the grade would be placed on the bulletin board for public inspection, whether or not the paper was listed, to discourage frivolous submissions to the panel. Reviewing would reimpose on the community some of the odious burden of refereeing, but it would be a much lighter load. Only a fraction of the papers would be submitted for grades, there would be no endless exchanges back and forth, and the reappearance of the same rejected paper at one journal after another would vanish with the vanishing of the journals themselves.

The fact is that journals are obsolete except as archival repositories, and even in this apparently benign role they waste such colossal amounts of shelving that plans are afoot to move them to compact disks. In the meantime we must either live with impossibly overcrowded libraries or devote construction funds better spent on laboratories or classrooms to library expansion.

Our failure to recognize the obsolescence of journals has restricted effective scientific communication to in-groups and cliques and is destroying our libraries. The sooner we get rid of journals, the better. Even the pleasure they once afforded of seeing a typed manuscript with handwritten equations transformed into beautiful fonts with justified margins has turned to ashes. If one has devoted any care at all to the preparation of a manuscript, the published version is now deeply disappointing, being visually less attractive and replete with errors introduced by the unnecessary process of retyping, or at least reformatting, the text to meet the journal's own specifications.

Why do we still live this way? What are we waiting for?

10. Postscript

1. I have almost always lagged behind in the computer revolution. I first heard the term "electronic mail" from my Cornell colleague Kenneth Wilson, an early prophet of Computopia. It sounded to me like one of his sillier enthusiasms. Why send written messages from one computer to another when the telephone provided direct mind-to-mind contact?[1] But in writing *Publishing in Computopia* I had a genuine vision of my own, producing a prophesy that was realized, with stunning precision, during the following quarter century.

   Three months after *Publishing in Computopia* appeared, Paul Ginsparg established just such a bulletin board on a computer in his office in Los Alamos. Originally based on email, it evolved into xxx.lanl.gov, and in 2001 it moved with Ginsparg to Cornell, where (as the next stage in a proud Cornell tradition that began with the founding of *Physical Review* in 1893) it is operated to this day, under the auspices of the Cornell University Library, as arXiv.org. Ginsparg had been thinking of such an undertaking before my column appeared, but reading my plea ("Why do we still live this way? What are we waiting for?") stirred him to action.[2]

2. In January 1992 *Physics Today* had six pages of letters responding to *Computopia*, almost all of them critical. It is interesting to

---

1   For other such moments with Ken Wilson, see Chapter 41.
2   More about the very early days of arXiv can be found in Chapter 12.

compare the many reasons offered for why my proposal would not work, with the actual operation of arXiv today.

(a) In the absence of refereeing authors will submit endless revisions, cluttering up the bulletin board.

▸ One can indeed submit revisions to arXiv. They are given consecutive version numbers. The default download is the latest version, but all earlier versions are retained and can be examined. Most people take submission to arXiv as seriously as submission to a journal. Most papers are never revised, and more than one revision is unusual.

(b) Any real computer system would probably be overloaded in a few weeks. Where will all the disks and tapes be stored? Who will retrieve them? What happens if fire or flood destroys them? What happens if the computer crashes or is closed for maintenance, or the phone lines go down?

▸ A charming failure to anticipate the prodigious growth in storage capacity, which has expanded enormously faster than the (impressive) rate at which use of arXiv has expanded. Inexpensive mirror sites around the world guard against local natural disasters.

(c) What about figures and diagrams? Will everybody have to have the same software?

▸ Yes. It's called Portable Document Format (pdf). It appeared the following year. A version that reads documents with figures and diagrams can be downloaded free of charge.

(d) The bulletin board will be vandalized by armies of hackers, who will attack the computers of anybody foolish enough to use it.

▸ This turned out not to be a problem for arXiv, though it has indeed been a major headache for the World Wide Web, which was just coming into existence at this time.

(e) Libraries will not have enough terminals to meet the demand. Scientists in less developed countries will not have access to terminals.

▶ A failure to anticipate the current ubiquity of increasingly inexpensive access.

(f) The bulletin board would be flooded with "hastily developed poorly written pieces that would not be accepted by any current journal." And "in preprints the authors can state anything they wish and very few scientists have the time to study or point out the numerous errors in the flood of preprints being distributed."

▶ These worries failed to anticipate the seriousness most people attach to posting their work on arXiv. It concentrates self-critical attention to know that your paper will be examined the following day by a significant fraction of all the interested people in the world, and that there will be no intermediate referee's report to give you a chance to correct major blunders. There is a higher level of self-refereeing on submissions to arXiv than on conventional journal submissions.

(g) "The entire chain of technologies needs to be faster and cheaper by an order of magnitude or more before electronic journals can become a viable alternative to paper ones."

▶ Indeed. Speed has gone up and cost has come down by many, many orders of magnitude. Everybody now reads journals online, from their homes, or from their offices, or wherever they happen to be using supercomputers carried in their pockets, called telephones.

11

What's wrong with those grants

My colleague Professor Mozart burst into my office, just back from a pro-SSC rally in Washington and still full of excitement. "The police estimated the crowd at seventy thousand, but it was at least a quarter of a million. It makes you proud to be a physicist. And to top it all off. while I was dodging tear-gas canisters, it came to me!"

"Tear gas, at a pro-SSC demonstration?" I gasped in disbelief.

"Yes," he confirmed. "An enormous crowd, unaware that they should have been addressing their concerns to Congress, started to march toward the White House chanting, 'Hey, hey, Allan Bromley, give us the Higgs or we won't go calmly!' and the Secret Service must have panicked. Those teenagers can be frightening, you know. They really get quite out of control when they think we might pass up an opportunity to find the Higgs. And those MIVeBs can be pretty alarming too, when they're on the move,"

"MIVeBs?" I inquired.

"Mothers for Intermediate Vector Bosons," he explained impatiently, unable to disguise his disdain at how out of touch I was with the Movement. "But then as the first canisters started to pop, I realized how simple the solution really was."

"Solution to what?"

"The funding problem for the individual investigator, of course! I can't imagine why nobody has thought of it before. We simply abolish all such grants, freeing the investigators to return to the full-time pursuit of their individual science." He settled into my only comfortable chair, beaming with satisfaction.

I've heard some zany things from Mozart before, but this one was just a little too self-serving to let pass. "Very fine for you, W. A.," I said with ill-concealed scorn, "who loathe writing proposals and progress reports and feel no responsibility for training the next generation of physicists. But what about the more conscientious members of our profession? How are they to keep the enterprise of small science alive?"

I hadn't intended to be so brusque with him, but it really is disgusting to see how much happier he's become since his grant was cut. Undeterred by my swipe, he continued.

"You don't understand. I'm not proposing to abolish support for small science—just to stop distributing it so irrationally. Take that next generation. Why do the agencies give out so few graduate student fellowships, and only for the first few years? Because they've tied up most of their funds for student support in individual-investigator grants. The consequences are appalling. Instead of doing their PhD research with the scientifically most congenial professors, students have to go where the money is or work without support. Absurdities abound. Our colleague Smetana, who flourishes with five or six students, has scarcely funding enough for one, while Beethoven, who works best without any, could easily support two or three. The best thing we could do for the next generation would be to remove student support from research grants and divert the funds into a greatly expanded program of predoctoral fellowships that would take promising students through their full graduate careers. Think of the benefits! Students could go to the professors who do the best science, which they're far better placed to judge than the reviewers of grant proposals. Professors would have more time to spend with students if they didn't have to forage about for their care and feeding. Furthermore, we could probably support half again as many students with what the agencies saved on indirect costs by distributing the funds directly to the students as fellowships ..."

"Hold on, W. A.!" I shouted. "You can't strangle the universities like that. Somebody has to pay for those indirect costs."

"… and of course the same goes for postdoctoral support," he concluded, oblivious to the note of realism I had tried to sound. "Now once we drop students and postdocs from the grants, is there any valid reason to have grants at all? Funds for travel?" he suggested, eyeing with disapproval the folders of airplane tickets in my in-basket. "Don't be silly! Everybody knows most people spend far too much time at conferences. Why? So they can give talks, preferably invited ones, and publish papers in the proceedings to fatten up their next grant application; so they can meet with like-minded colleagues to coordinate political action on the funding crisis for individual investigators; so they can have some relief from the day-to-day grind of writing, reviewing, and reporting on proposals. Publication costs?" he went on, glancing uneasily at the two-foot stack of unopened journals threatening to topple over on my desk. "But you yourself just made the case brilliantly[1] that journals are obsolete and should be phased out. Long-distance phone calls? Postage? Copying costs? Faxing? Most employers cover such employee expenses routinely, and it's high time the universities did too."

"There you go again, dumping more costs on the universities!"

"The only legitimate item for a research proposal," he continued, seemingly indifferent to the plight of the universities, "is the direct material cost of the research itself: capital equipment, stockroom goods and the relevant physical plant. If we retained individual-investigator grants just for those items, then theorists would only have to write occasional proposals for computers and software. Experimentalists would have to work harder, as they do now anyway, but only to get the equipment and supplies directly required for their experiments. And many of these are the kinds of expenses on which indirect costs aren't even charged—more savings for the agencies!"

1 Chapter 10.

"But ... but ... but ..."

"Of course," he intoned solemnly, "there would be no more summer salaries," adding brightly, "just as there didn't used to be 30 or 40 years ago, when the whole system veered off down the wrong track. In those days science professors weren't any different from the others. They were paid for 12 months, and got 3 summer months off to refresh their intellectual powers. The pay wasn't wonderful, but that summer of freedom was worth a lot. Since then the fiction has grown up that we're only paid for 9 months, requiring an additional two to three ninths of our annual salary to recompense us for the sacrifice of not spending 3 solid summer months on camping trips, beaches, or round-the-world voyages. Can anybody believe that? Why, there's nothing we'd rather do than spend the summer working in our laboratories or at our computers and desks, without the distractions of the academic year."

"You entirely miss the point," I chided him. "If summer salary is abolished there will be a massive flight of present and potential academic scientists into industry,"

"Wonderful! Technology transfer is what it's all about, and what better way to strengthen the links between fundamental science and economic competitiveness?"

"Not if it leads to the collapse of academic science in America."

"Are you trying to tell me that our best academic scientists are in the universities for the money? Be serious! There's no shortage of people to populate political science departments, though they could make far more in the legal profession. How do we keep our economics faculties when a fresh MBA gets more than a senior professor? The fact is, we're paid to do what we enjoy most. There won't be any exodus. And if there is a real problem of equity—if academic salaries fall so far behind that the professoriate has to take vows of poverty—then it's the responsibility of the universities ..."

"Hold it!" I shouted, determined to stop him at last, "First you take from the universities all recompense for their contributions to the indirect costs of sponsored research: then, having deepened their fiscal crisis, you blithely announce that if summer salary is required to stave off the wolves, the universities should supply it themselves. But tuition is already at unacceptably high levels. Your scheme relies on an economic miracle ... hopelessly naive ... how innocent can you get ... overdose of tear gas ..." I sputtered on furiously.

"Are you running for provost?" he inquired politely. "Since it's obviously in the national interest to keep science strong in America, the Federal government should directly and routinely reimburse the universities for the costs of excellent research beyond what tuition payments can legitimately support. And what better way to start rethinking how to do it than by abolishing all the individual-investigator grants, where the overhead tax is at its most bizarre. A reimbursement scheme should be rational, but the present one is, as Jimmy Carter said of the tax code, a disgrace to the human race. Every university negotiates its own formula, whose relation to actual indirect costs, a concept more shrouded in mystery than the true nature of the quantum state, ranges from the unavoidably obscure to the explicitly ludicrous. Bonanzas for a university, like the donation of a new research building, can turn into disasters for every scientist on campus, in the form of more points on the overhead rate. The system pits scientists against administrators in battles of increasing ferocity, even though their actual interests are virtually the same. Growing hordes of academic bureaucrats are required simply to monitor the process, whose salaries drive indirect costs higher still.

"Orgies of irrationality result from tying this reimbursement to ritualistic formulas and collecting it from individual investigators as an across-the-board tax. This by itself is reason enough to abolish such grants. Since tuition revenues alone can't support vital scientific studies, the government should institute a direct system of research subsidies to universities. The support could be phased in at the level of what the universities are currently receiving in indirect-cost recovery from all their obsolete individual-investigator grants. Future payments would be updated every ten years at the recommendation of national panels of peer reviewers similar to those that evaluate proposals for major research centers. The evaluation would be based on the total research accomplishment of the university in the previous decade, and the amount would be entirely decoupled from the content of whatever individual grants remained, so that the entrepreneurial successes of one could no longer impose ridiculous taxes on the funds of others. A basic allotment for equipment and supplies would also be awarded directly to the universities for internal distribution by local decision, which is invariably better informed than the opinions of outsiders. And with the decennial reviews in mind, the universities would be far better motivated to see that the funds are well spent. Only people with special needs for extraordinary equipment would have to apply directly to the agencies for individual-investigator support. Besides paying for traditional indirect costs, universities could use the funds, at their discretion, to resupply scientists with some of the 'expendables'—telephone calls, I would hope, at a minimum—that are currently, and absurdly, covered by research grants."

And he beamed at me. I threw him out of my office. "Go join the MIVeBs!" I shouted after him. "They could use some better chants! I've got more important things to do." And I did: a progress report to write, a new proposal to submit, and three to referee.

11. Postscript

1. Professor Mozart, whose passion for the Superconducting Super Collider (SSC) was confirmed in a letter written while fund-raising in Kazan,[2] here enthuses over how the project has captured the imagination of the teenagers and mothers of America. Allusions to contemporary figures and to the

2 See the Postscript to Chapter 9.

Vietnam-war demonstrations two decades earlier may escape today's teenagers and even their mothers. Alan Bromley, a distinguished physicist from Yale, was the science advisor of the first President Bush. Both Bush the First and Bromley supported the SSC, which was killed by Congress during the early days of the subsequent Clinton administration.

2. The mob mistakenly marches on the White House only because I couldn't resist the rhyme *Bromley-calmly*. Their chant echoes the lines addressed to President Lyndon Baines Johnson in the demonstrations of the later 1960s: "Hey, hey, LBJ, how many kids have you killed today?"

12

What's wrong in Computopia

Professor Mozart burst into my office, waving the January 1992 issue of *Physics Today*. "What are you doing here, W. A.?" I greeted him in surprise. "I thought you were abroad fund-raising for the SSC!"

"Just got back," he gasped, having apparently run up all five flights of stairs. "Castro says he'll provide all the cigars if we can persuade Bush to lift the sugar quota. Just sent Bromley a memo. Don't see how Congress can drag its feet any longer—especially when we remind them that accelerator physics gave us ride-on lawn mowers, sliced bread, and the compact disc. But what about this response to your call last May[1] for the abolition of journals in favor of electronic bulletin boards? Ten letters to the editor—all but two hostile? As a pundit, you've got it made!"

"Thank you," I replied sourly, "but the fact is I received even more letters that were wildly enthusiastic—by far the biggest response I've ever had."

"Don't tell me," he said, lighting up an enormous Havana. "All the favorable correspondence came by email. No copies to *Physics Today*. Shun the print media. Matter of principle."

"You've got it," I confirmed, suppressing a gasp myself. "My supporters are all children of the network. I doubt they even use the telephone anymore, except as ancillary to a modem. They want me to lead the way into the shining electronic future, writing software,

1 Chapter 10.

designing hardware, lobbying professional societies, organizing boycotts, raising funds ..."

"Leave the fund-raising to me," he ordered through the smoke. "Your immediate problem is to answer your critics. How could you have expected to attack the refereeing process and come out unscathed? Don't you realize most people can't write an acceptable laundry list without peer review? Without referees we'd soon be promulgating inchoate blather. Can you imagine what *Hamlet* must have looked like the first time Shakespeare submitted it? Why, somebody once told me that *Othello* is what *Titus Andronicus* turned into after half a dozen exchanges. And you want to abolish refereeing!"

"Never mind how peer review operates under the current system," I interrupted. "What none of the critics have noticed is how much *better* it will work in Computopia."

"No doubt you're thinking," he murmured through the fog, eyes half closed, "of a parallel bulletin board of criticisms and errata."

"Precisely. Those genuinely interested in any paper—namely people who, unlike today's referees, spontaneously choose to read it—would have the opportunity to post laudatory or critical comments for the benefit of subsequent readers; the author, of course, could post a reply. The system would make available upon request the comments currently on file for each document."

"Every paper its own seminar talk!" Mozart burbled enthusiastically.

"Not quite," I pointed out. "Only those interesting enough to elicit a response. Can you imagine it—every paper of note collecting a constellation of signed commentaries from interested experts, available to all? What could be more enlivening! But what really surprised me was that nobody on either side of the issue seemed interested in the problem that led to my proposal in the first place."

"You mean the undemocratic monopolization of cutting-edge science by self-selected cliques through the proliferation of preprints as the primary publication procedure?" he asked, smiling with approval at the train of perfect little smoke rings that emerged from his preponderance of p's.

"Precisely," I coughed. "Journals or no journals, nobody is going to stop the circulation of unrefereed preprints. Most fields of physics have been exchanging their most important communications through preprints for well over a decade. Posting such documents on publicly accessible bulletin boards is hardly a utopian vision—it's a moral imperative. Furthermore, it works. The string theory and two-dimensional gravity people have been doing it for some time now—almost a thousand of them, worldwide. Nobody in the field sends out paper preprints anymore. Everything is fully automated—the system runs itself. Anybody anywhere can subscribe by sending in a single email message. You get a daily list of new titles and abstracts, and can call up the complete text of any paper that might interest you in a matter of minutes. Papers are available for about a year and the whole thing occupies about 10% of the hard disk of one workstation in the office of one physicist at Los Alamos, operating in the background with a negligible drain on CPU time. There are no frivolous submissions and no practical jokes, just a large number of serious people exchanging ideas."

"String theorists not utopian?" he twinkled through the fog. "You've got to be pulling my leg."

"I know it's hard to believe," I said, "but this could well end up as their greatest contribution to science. They have seen the short-term future and converted their vision into a practical scheme for propagating their thoughts about the physics we may be using in the ultralong-term future. It's totally democratic: everybody can be as up to date as the leaders of the field, without having to get onto anybody's preprint lists. Indeed, those in faraway places who were on the lists now benefit from the elimination of the old boat-mail delays in delivery. Everybody is better off. Why, even Glashow uses it!"

"Completely up to date on the physics of the 520th century," Mozart, mused. His devotion to the SSC has led to an uncharacteristic intolerance of investigations that probe beyond the TeV range. "Considering when the relevant experiments are likely to be done, falling a century or two behind in the literature would hardly seem to be a handicap."

"You miss the point, W. A. It's a great intellectual adventure. Before the bulletin board opened up the field, you had to know the right people to get into the game. Now everybody can play."

"All well and good for a dedicated bunch of fanatics," he snarled through the haze, "but what happens when you try to extend that to a serious field like superconductivity, where you might end up with ten thousand subscribers?"

"You've been off fund-raising too long," I chided him. "Suppose every single subscriber also submits four papers a year, each 50 kilobytes long—twice the length of a *Physical Review* letter. That's 2 gigabytes a year. Why, even laptops these days can handle a tenth of a gigabyte. Do you really believe that existing technology can't give us a scheme that is capacious, inexpensive, easily accessible, capable of dealing with figures, and secure against accidents or deliberate sabotage? The files could readily be made available to libraries in a variety of inexpensive permanent storage devices at regular intervals for archival purposes, at a minute fraction of the cost of acquiring the same information on printed pages, and a minuscule fraction of the cost in floor space.

"But even if it strained us to manage it today," I pressed on, "if you consider where we are now and where we were a mere ten years ago, can you seriously doubt that in another ten years such technological problems as might currently stand in the way will have completely evaporated? One doesn't often have the chance to contemplate Utopia a mere ten years down the road; if we want to be ready we'd better act now. Let's face it, the real objections to such a scheme aren't technological."

He looked at me with uncharacteristic admiration, "You're right, technological objections are entirely beside the point. If the scheme can't be shown to be inherently objectionable under ideal technological conditions, then it's inevitable." And as I leaped unsuccessfully to stop him, he ground out his cigar on a loose issue of *Physical Review Letters*. "The real trouble is lack of closure. When your paper appears in a journal, that's it. Your thoughts on the subject at that moment are frozen into the archives. Awareness of this sharpens the attention. It enforces a level of self-criticism, thoroughness, and just plain careful proofreading that would simply not be elicited if you could repeatedly ship communications off to the bulletin board at a moment's notice. That's what you're giving up. It's not worth it." He brushed the sparks onto my rug.

"But it's a sociological fact that this simply hasn't happened in the existing schemes," I protested, frantically stamping out the glowing embers. "And there are powerful reasons why it won't happen. Science may once have consisted of discovering the truth and making it available to others, but today there's another problem almost as difficult: getting anybody to pay attention. When mountains of new work appear each month, most of it vanishes unnoticed. People on the bulletin board who acquired a reputation for repeated resubmissions of trivial revisions of earlier manuscripts would rapidly lose any audience they might once have had. They would cause nobody any further trouble, and the disk space wasted with their unread offerings would be vastly less valuable than the space they currently waste in the libraries, playing exactly the same game with conference proceedings. Since everybody could easily keep track of the number of people who requested copies of their papers, perpetrators of trivial resubmissions would quickly learn that nobody was paying any attention."

"That's right," said Mozart. "Give everybody a list of the names of all the readers of their papers, so they can go around harassing them or making paranoid accusations of plagiarism."

"The question of whether to preserve the anonymity of requesters is precisely the kind of interesting and important issue people should be discussing today, in the decade before Computopia sets in, rather than wasting their time declaring its impracticality."

"Of course," said Mozart, firing up, to my horror, an even bigger Havana, "your scheme is death to browsing. There's simply no way for a computer to simulate the experience of cuddling back in a comfortable chair and taking a leisurely literary stroll through a pile of promising periodicals in search of something entertaining but completely unanticipated." Three tiny smoke rings punctuated this thought.

"True," I admitted sadly. "The abolition of the library card catalog has already deprived us of a similar pleasure. All progress has its price."

"But possibilities do come to mind," Mozart offered cheerily. "Browsing is now necessarily an entirely random business, since it's humanly impossible to sample more than a minute fraction of the literature. One could easily build into the bulletin board a browsing capability that would present the determined browser with a dozen titles, randomly selected from a prespecified set of areas, defined as broadly or narrowly as required. You could even request a collection of randomly selected single pages from those within specified browsing parameters. Nothing would be missing but the easy chair."

"But the easy chair is important," I insisted. "So is the feel of the paper, the smell of the ink, the crackle of the glue, the rustle of the pages, and the pungency of the mildew. A humane Computopia will have to maintain a small number of browsing periodicals."

"Peer reviewed?" he shot back.

"Of course," I conceded. "But since a primary criterion for acceptance in a browsing magazine would be readability, as soon as the task of refereeing verged on the disagreeable that in itself would be *prima facie* grounds for rejecting the piece. 'This paper should not be published because I find it tiresome' would be an entirely satisfactory report, and no great burden would be imposed on the refereeing community." Mozart rose from his chair, wreathed in blue-white vapor. "Our colleague Schubert maintains that the real problem is human vanity. He says the present system hasn't died because we still believe that there is glory to be had in getting our words onto the printed page. Hard to understand, given the evidence that most printed pages languish unexamined, but maybe it's enough for the author alone to see the finished product—a sort of papyro-narcissism. Perhaps," he added, as he and the fumes drifted out my door and I dashed, wheezing, to the window and threw it open, "we could furnish such people with dummy journals where their papers were embedded in randomly selected collections of writings of great distinction. We could sell these volumes to such authors at a profit." The door swung shut behind him, but I could just discern his final muffled thought as he rambled off down the corridor: "Send the proceeds straight to Waxahachie. Pay for all the ashtrays."

12. Postscript

1. Professor Mozart's efforts to generate international financial support for the Superconducting Super Collider (SSC) have led him to Havana, revealing his taste for fine cigars. But his primary purpose is to discuss some of the responses to *Publishing in Computopia*, Chapter 10. In the course of our conversation I have more to say about the Los Alamos archive, which I only learned about after *Publishing in Computopia* had appeared.

2. Practitioners of string theory are teased by both of us. String theorists attempt to understand elementary particles by positing new geometrical structure at length scales a trillion million times smaller than the size of an atomic nucleus. Professor Mozart, who expressed astonishment that quantum mechanics

continues to work at length scales a mere million times smaller than those for which it was devised, regards such efforts as utopian.
My remark to Mozart that the e-print archive (now arXiv. org) "could well end up as [string theory's] greatest contribution to science" achieved a degree of notoriety. I still believe it.

3. The capacity of storage devices on a laptop in 2015 is thousands of times greater than the tenth of a gigabyte I mention approvingly in this 1992 essay.

13

What's wrong with those talks

My friend Professor Mozart recently ran across some advice to young physicists on how to give talks.[1] He came to me seething with indignation. "What's the problem, W. A.?" I asked. "I thought Jim Garland spelled out concisely and effectively just about everything the novice ought to take into consideration."

"As you say," he snarled, "it was a precise recipe for how to produce a contemporary physics talk—an almost perfect codification of all the ingredients."

"Well what more could you ask?"

He gave me a look of withering scorn. "The contemporary physics talk is a disaster," he proclaimed. "The only pleasure it affords is the relief that washes over you as you realize, finally, that perhaps the end is in sight. To assemble a respectable audience you have to bribe people with cookies and muffins. You must offer gallons of coffee to those honorable enough not to take the food and run, to help them maintain consciousness during the next hour. The article in *Physics Today* did a masterful job of passing on to future generations everything necessary to maintain this dreary art form."

"You're unfair," I reprimanded him. "There are too many things about lecturing that you, an experienced speaker, simply take for granted. If you think the article gave young physicists bad advice, have you anything better to offer?"

1 James C. Garland's article in *Physics Today*, July 1991, page 42.

"They were not given bad advice. They were given excellent advice for making the best of an inherently hopeless situation. But pretending that the standard physics talk of today is an acceptable form of communication breeds hypocrisy in the old and experienced, and nurtures self-doubt in the young and innocent, who not only have to undergo the wretched experience of attending physics talks but also torture themselves worrying why they're not enjoying the ordeal. I would have urged speakers to get to the root of the problem."

"And just what might that be?"

Without another word he thrust into my hands a battered handwritten manuscript covered with coffee stains and smeared with muffin crumbs, evidently labored over during many hours of intolerably dull seminars and colloquia. Then he walked off in a huff.

Though appalled by some of the opinions expressed in the document he handed me, I reproduce it below in its entirety as a counterbalance to the conventional wisdom.

Advice to beginning physics speakers (and intermediate or advanced ones)

William A. Mozart*

If you have taught physics you know it is virtually impossible to write too easy an exam. Yet nobody acknowledges that the same is even more true of the physics talk. It is absolutely impossible to give too elementary a physics talk. Every talk I have ever attended in four decades of lecture-going has been too hard. There is therefore no point in advising you to make your talk clear and comprehensible.

---

* **Bill Mozart** is Rachmaninoff Professor of Physical Science somewhere in the depths of central New York. He has been forced to embed these precepts in another's article, because *Physics Today* discriminates against imaginary people.

You should merely strive to place as far as possible from the beginning the grim moment when more than 90% of your audience is able to make sense of less than 10% of anything you say. It is in the nature of physics talks that they should be boring and confusing. You, the speaker, struggled through ten years of college and graduate school to reach the point where you could do research in your chosen area, acquiring arcane skills available to only a narrow range of practitioners. To attempt in the space of an hour to provide your audience with even the minimal background necessary to savor your recent research achievements is a doomed undertaking.

Yet we do give talks. Why? Only when this is understood can there be hope of producing an acceptable lecture.

The best reason to lecture on your work is that it affords you the opportunity to rediscover why you did it. The most important question to ask yourself in preparing your talk is why on earth any physicist might be interested. This is dangerous: there is always the risk you will find no answer. But that is not necessarily a cause for alarm. Often when working on a problem for a long time, one does indeed forget what first led one into that line of endeavor, so if at first you can find no answer, think some more. What is there in the subject to capture the imagination of one lacking your highly specialized skills?

Give yourself a week. If you still can find no reason why anyone not directly involved in the work should find it anything but tediously obscure, then you should find something else to talk about. Indeed you might then seriously consider finding another area of research. Often merely preparing to give a talk can yield up such beneficial insights without your ever actually having to give it.

But suppose you do remember why you got into your current line of research. If you succeed in conveying that early freshness and excitement to somebody else, your talk will be an unqualified success, even if you never manage to describe a single one of the splendid things you uncovered when the project was well under way. Those interested in such technical matters will ask you questions in private. For no matter how detailed you might be tempted to make your talk, it cannot possibly be detailed enough for those few who are knowledgeable enough to appreciate such refinements. And no matter how basic and elementary you make your treatment of those fascinating technical accomplishments, virtually none of them will penetrate the minds of the overwhelming majority of your audience. Your only goal must be to furnish ordinary physicists with some modest glimpse of what sustains your own interest in your subject.

What brings even well-intentioned efforts to grief is the misconception that it is necessary for speakers to talk about their own contributions. There is no need to say anything whatever about what you did yourself. Your personal work in the field qualifies you to give a talk only because it may have led you to discover how to break through the formidable barriers preventing the subject from engaging the interest of outsiders. If you can manage to do this *and* encompass a contribution or two of your own, that is fine. But if your own contributions are unfit for public display in such a forum, that too is fine, provided you do not persist in displaying them anyway. This should be kept in mind even when designing "job talks" or presentations at specialized conference sessions. Sometimes you have no choice but to speak of your own work, but even then it is best to devote the greater part of your talk to giving the clearest possible context for that contribution.

Never, ever, have I heard anybody complain about a talk on the grounds that "I understood everything in it." People feel good after talks they understand. Even those few people who hear nothing they didn't already know can derive substantial enjoyment from hearing their subject presented well. The most important thing your talk can do for such experts is to give them an opportunity to learn how to do better in their own talks.

Other points to keep in mind:

▶ Humanists, who take words more seriously than physicists do, often read their talks from a prepared text. When the talk is delivered with animation and impromptu asides, the results can be spectacular, for the written language is more powerful and concise than informal speech, and a richer and more attractive medium. Most physicists deem it undignified or unsporting to read a prepared text. Rubbish!

▶ The physics talk has, in any event, evolved toward the reading of a prepared text, but in an entirely unsatisfactory way. Many physicists do read their talks, not from a paper text, but from a sheet of transparent plastic projected on a screen. This combines the worst of both approaches: The spontaneity of improvisation is lost, but the elegance of writing is not achieved, since the verbal contents of the plastic sheet are fragmentary stammerings, not written language. To make things worse, text on plastic sheets can be read by an audience faster than the speaker can anticlimactically deliver it, unless the abominable practice is employed of covering up most of the plastic until the moment of revelation. Sheets of plastic must never be used to convey the purely verbal, which should be either spoken extempore or read aloud from a paper text.

▶ Sheets of plastic are only for illustrative figures, graphs, or data, and unavoidable elementary mathematical analysis in the absence of a blackboard. Even when so used they almost always have too much on them. Many in your audience will have an unobstructed view of only the upper half of the screen, and many will be seated quite far from it. You must therefore put very little on each sheet, leave the lower half empty and make everything extremely large and uncluttered. If your analysis or diagram is too intricate to present in this way, it is too intricate to be in a talk at all. Just as one should go through a manuscript many times, ruthlessly cutting the redundant, so too should one keep redesigning a plastic sheet to reduce its contents to the bare minimum. You will be present when the sheet is on display. Most details are better supplied orally.

▶ We are fortunate to live in an age of informal dress. When giving a talk, wear whatever makes you comfortable, remembering only that a filthy or outlandish costume may be viewed by your audience as a sign of disrespect or incipient lunacy. Do not worry whether all your buttons are buttoned. Once you start down that perilous path you can wonder whether there is ketchup on your nose, a large chalky smudge on your back, or a piece of stickum with a coarse message maliciously affixed to an inaccessible part of your person. Assume that if you are in disrepair somebody in your audience will have the kindness to call it discreetly to your attention, permitting you to fix the problem on the spot. If it's not called to your attention, it's not a problem. If it is, simply say, "Ah, mustard on my ear? Sorry about that," wipe it off and continue.

▶ On those few occasions when a physics talk delves into the history, sociology, or social psychology of the subject, the audience wakes up and listens. Though most professional journals frown on such digressions, they are entirely appropriate in a lecture. Reading aloud from the reports of hostile referees, for example, almost invariably rouses an audience from its stupor as well as giving you a rare opportunity to make it vividly and painlessly aware of your own contributions.

▶ The ubiquitous heavy-handed concluding summary should be omitted; a talk should tell such a good story that a summary is uncalled for. Imagine *War and Peace* ending with a summary. There is no better way to make an audience happy than briskly finishing a talk five minutes earlier than it expected you to. Like this.

13. Postscript

1. James Garland, whose article advising young physicists how to give talks inspired Professor Mozart to produce his own advice,

was a graduate student at Cornell when I arrived there in 1964 as an Assistant Professor. While I was Director of Cornell's Laboratory of Atomic and Solid State Physics, Garland was Chairman of the Physics Department at the Ohio State University, and in 1987 he lured away from us one of the most important members of our laboratory. That man, in turn, lured to Ohio State the most important member of our department. So we've been in touch over the years.

When Garland published in *Physics Today* advice to beginning physicists on how to give scientific lectures, I felt he was advising young people how to give concerts on an instrument that was hopelessly out of tune. I worried how to say so politely. Another job for Professor Mozart! He could boorishly condemn both Garland and one of the most important institutions in the life of most contemporary physicists, while I valiantly defended the status quo.

2. The "sheet of transparent plastic projected on a screen" was the dominant mode of giving a physics talk in the second half of the 20th century. It has gone the way of carbon paper and the typewriter. But all of Bill Mozart's advice for such "transparencies" applies equally well, if not more so, to PowerPoint presentations. The temptation to clutter PowerPoint pages with irrelevant or marginally relevant material seems irresistible, even though each page is shown so quickly that nobody can pay even superficial attention to all of its contents.

14

Two lectures on the wave-particle duality

During the recent[1] presidential election, I dreamt that two of the candidates had concluded from interviews with focus groups that there might be some anxiety among the American public over the foundations of quantum mechanics. Concerned that by homing in on so esoteric a topic they could lose the attention of the people, the two men had made a direct assessment of public interest by quietly employing their rhetorical skills in unpublicized lectures at local events such as church barbecues, farmers' markets, or demolition derbies. I could never learn soon enough about these performances, always arriving just as a lecture ended. By conducting exit interviews, however, I managed to put together fragmentary transcripts of what took place, which were so vivid that I was able to jot them down in the morning.

In a subsequent dream I read these texts back to my interviewees, who agreed that although I had failed to capture the full brilliance of the argumentation, I had at least succeeded in conveying the flavor of the insight these remarkable men brought to the problems that have puzzled and delighted physicists for so many years.

The candidates' experiments were not a success. Both men concluded that the time was not ripe to bring these great issues before the public. Indeed, in my third and final dream I was forced to endure an interminable postelection analysis on public TV, in which the panelists concluded that by distracting the

---

1 That of 1992.

two candidates from more pressing issues, their love of quantum mechanics had contributed significantly to their defeat. I'm sure there are lessons for physicists from this cautionary tale, but I offer here only the texts of the lectures themselves, which I believe form an important chapter in the intellectual history of our times.

The first lecture

Now it's really very simple, OK? Over here's an electron, moving toward this wall, kind of like a cur dog slinking toward his kennel. Only there are two doors to the kennel, like the two doors in the wall here in Figure 1.

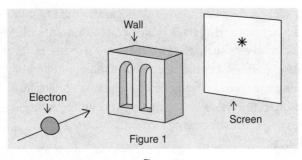

Figure 1

Now, over here on the other side of the wall's a screen. Now then, the point is, the electron ends up making a mark on the screen, kind of like a fly makes a speck on a kitchen window? So the electron starts over here on the near side of the wall, and ends over here on the far side, in this little flyspeck. Now you and I, we ask, "How did that electron get from over here to over there?" They ought to be able to give us a straight answer to that question, right?
    Wrong!
    You and I, we ask, "Did that electron go through this hole or did it go through that hole?" and I find it fascinating that they will not give us a straight answer. They say, "That isn't a proper question." If

you were in business and gave an answer like that they'd laugh you right out of the boardroom.

Now, of course it's a proper question. We know perfectly well it's a proper question. But! They go around telling sensible folks like you and me that it isn't a proper question. Why are they doing that? Any child knows why. Point is, they don't know the answer! I rest my case.

Now then! They simply don't want to find out which door the electron went through! Isn't it just fascinating? They've been telling us this since nineteen hundred and twenty-five. For nearly 70 years they haven't been able to figure it out. They're still ducking that question! They'll do *anything* to avoid answering, making up tales about invisible waves and things not really being there that would have got them a whipping if they'd told them to their mothers when they were little. Now, when you get to be the boss here's what we're going to do.

It's just this easy. We are going to get the best minds together—world-class minds. And we'll say to them, "Just look at this—we've got an electron, we've got a wall with two doors, and we've got a screen on the other side of the wall." We'll tell them: "Boys, you just roll up your sleeves, don't be afraid to get your hands dirty, get in under that hood, and you watch that electron really carefully. Then you go and you have a raging debate about which door that electron went through, and when you've heard all the arguments and made all the points, then after that no-holds-barred discussion, you come out with a proposal; door 1 or door 2."

Then we have a town meeting. We bring that proposal to all of you and see what you think of it. If you've got a better idea, you just let us know. If it sounds like it'll work, we'll tear everything up and give it a try. But we have got to sit down and make up our minds and get on with the important stuff.

Now if you want more talk about not this door and not that door, both doors, no doors, improper questions, invisible waves, messing things up by poking around, and all that kind of slow dancing, then you don't need me and I've got better things to do with my time. But! If you want to answer this question once and for all, then I'm your man, and we'll get to work and settle the whole business. I'm not saying it's going to be easy—of course it won't, but we'll have a lot of fun together clearing it up. And when we've straightened it out all those boys in their blow-dried hair and thousand-dollar suits will come and say, "Well, you called our bluff, but we love you anyway." Fact is, they'll be tickled pink to know what door the electron went through!

I adore those quantum mechanics folks. They're fine patriotic citizens. But they have just got to stop denying what you and I can plainly see. Wouldn't it just be stunning if we got them together when we were done and they turned to us and just said, "Yes, you're right: It went through *that* door"? I rest my case!

The second lecture

I'll come right to the point. When they asked—when they said to him, "Governor, is the electron a wave or is it a particle?" the Governor said, and I'm quoting him here—he can't waffle out of it this time—this is what he said, and I quote: "It's neither; the concepts of wave and particle as we ordinarily understand them simply don't apply to the electron." Simply don't apply! That's what he said! And this is a man who wants to be president of the United States of America!

Now, let me tell you something. There's nothing new about this kind of talk. These are old, worn-out ideas. They originated over 75 years ago. In Europe! I'm not one to question anybody's patriotism and I really don't think patriotism has anything to do with quantum mechanics, *if you stop doing it at the water's edge.* But when you go off to a foreign capital—London, Moscow—he says he doesn't remember who he met there—and you organize a group of people to go around saying it's neither a wave nor a particle—well I guess I'm just old-fashioned enough to think that it does matter which it is and I think the American people want straight talk about which it is.

I don't know about these two bozos: Big Bore and Uncertainty Man. Seems to me he was pretty certain about what he was up to back in the war, Mr. Un Certainty. Not one to question him myself, but I don't know about all this Un Certainty stuff. Not the kind of values I was brought up with, I can tell you that. Un Certainty! Talk about uncertainty, in Montana the Governor says, "Oh, yes, sure, it's a particle." Then, *one day later*, he's in Alabama and he's saying, "Absolutely, it's a wave." Wait!—wait, it gets worse. He said—and this is in writing, he can't waffle out of this one—if he becomes president, you know—the Ronald Reagan High Energy Physics Center—you know what they're going to call it? The Waffletron!!—but don't worry, it's not going to happen. Just the other day, he said, "Well, I guess on the whole I'd say it's a particle but there are circumstances under which it surely does behave like a wave."

Well Governor, maybe you can get away with that kind of talk in a rotten little state like Arkansas but we're talking big-league now. This is Oval Office stuff, buck-stops-here kind of thing. When that red phone rings at 4 in the morning … general at the other end, somewhere out in that desert … SCUDs flying over … half a million troops … world's oil supply on the line … and you say to him, "Well general, fact is it's neither a wave nor a particle but something else." … No way Josephine—buck stops!

You know, when we talk about family values, we mean knowing—really knowing—which hole the electron went through. It's the kids I'm thinking about. Call me old-fashioned, but those kids have to feel that, yes, their parents know where the electron is. That's what we mean when we talk about trust, about character.

Think about this. Suppose you turn on the television someday and the announcer is saying, There is a major crisis—an infrared catastrophe, out there in the desert, or, yes, an ultraviolet divergence in one of our cities, right here at home. Ask yourself this: Who do you want behind that desk at that time—a man who can't decide whether it's a wave or a particle, or a man who can make those tough calls?

Thank you, thank you very much, and God bless Determinism.

14. Postscript

1. *Two Lectures* was written shortly after Bill Clinton defeated George (the First) Bush's 1992 effort to be re-elected president, in a strange three-way election involving Ross Perot, an eccentric Texas billionaire, who received almost 19% of the vote, and may well have been responsible for Clinton's success. My first lecture is in the oratorical style of Perot; my second imitates Bush the First.

2. I've been asked why I left out Clinton, whose style and manner of speaking were at least as susceptible to parody as the other two. The answer is that I supported Clinton, and it is much easier to see ludicrous absurdities in the arguments of those you disagree with. I was to experience Clinton's absurdities in abundance over the eight years that followed.

3. The "Ronald Reagan High Energy Physics Center" was the name that the laboratory in Waxahachie, Texas would have been given, had the Superconducting Super Collider actually been finished, honoring the president whose enthusiastic support launched the project. President Clinton's lackluster support probably played a role in its demise.[2]

2 More about the Superconducting Super Collider in note 3 of the Postscript to Chapter 9.

15

A quarrel we can settle

Now that the standard model has been with us long enough to have become part of the commonplace wisdom of schoolchildren, it is high time to face head on the contentious issue of how properly to pronounce the word *quark*. Although only a condensed-matter theorist, I am proud to contribute what follows as a low-cost contribution to straightening out one of the annoying loose ends, others of which may have to wait considerably longer for their resolution.

As a rule, only native speakers of English hold passionate opinions on how *quark* is to be pronounced, and that is as it should be. One of the glories of the English tongue is that its comprehensibility is undiminished and its beauty even enhanced by the systematic mispronunciation of vowel sounds by non-native speakers. But it is a sad and ugly business when the wrong sounds emerge sporadically from the mouths of natives. Quite aside from such aesthetic considerations, it is surely the duty of us native speakers of English to set appropriate standards that the others, if they so desire, may strive to attain. Nowhere is there more need for clarification than in the hotly disputed case of *quark*.

The opinion of the majority is clear: *Quark* is pronounced to rhyme with *pork*. There is, however, a vocal and embittered minority, biting in its rejection of the prevailing view, whose members insist on pronouncing the word to rhyme with *park*. It is often argued in defense of this practice that the word is taken from the German name for a horrible yogurt-like fluid, the proper pronunciation of which unquestionably comes closer to rhyming with the

English *park* than it does with *pork*. This minority argument, however, is entirely spurious. If one were to adopt the German pronunciation consistently one could indeed rhyme *quark* with *park*, but only at the price of having to say *kvark*, which no native speaker of English has ever been known to advocate.

Clearly the decision must be made from a study of English usage and, I would maintain, either on the basis of the Irish variety of English, out of respect for the man who imported the word from German, or on the basis of the American variety, in deference to the man who transported it into physics. The fact that the international enterprise of physics is, on the whole (and alas for it!), not conducted in the Irish version of English and the fact that this column is appearing in an American magazine decisively tip the balance toward the latter approach, though I would be delighted to learn the views of anyone possessing (as I do not) a fine command of the Irish vowel.

For many years it was obvious to me that the majority view was correct, on the following grounds: Search through the dictionary for any word containing *ar* preceded by a *w* sound (including the *w* sound in *wh* or *qu*). Invariably you will find that such a word rhymes with *pork*. I cite *quart, wart, ward, war, warp, warm, dwarf, wharf, quarter, thwart*… To be sure, if the *r* is doubled there is a possibility of variation, as in *quarrel* or *quarry*, but even there the variation exists only as an alternative to the undiluted *pork* sound. (There are some truly marvelous things to be said about *arr* words, which a *Reference Frame* column is too brief to contain. I merely note that the methodology developed below is powerful enough to deal with them, lending further support to the conclusions reached here.) So we have a rule that I name after its simplest example:

The war *rule:* In American English *ar* preceded by any kind of *w* sound is invariably pronounced like the *or* in *pork*.

I lived happily under the *war* rule, quarking away with the best of them, always rhyming it with *pork*, for many years. Then one day I was rudely awakened from my customary lecture-room stupor by a speaker of the *park* persuasion who suddenly abandoned his transparencies to launch a scornful attack on the practice of pronouncing *quark* to rhyme with *pork*. Instantly awake and in full possession of my rested wits, I began to conjure up an enormous list of *war* words with which to demolish him during the question period, when a shocking thought entered my head:

Search through the dictionary for any word containing *ar* followed by a *k* sound, and it invariably rhymes with *park*.

How could I have failed to notice this for so long? In addition to *park* itself we have the bare *ark*, along with *bark, hark, lark, mark, embark, spark, stark, shark, snark* … I had clearly overlooked:

The ark *rule:* In American English *ar* followed by any kind of *k* sound is invariably pronounced like the *ar* in *park*.

Cognitive dissonance! How is one to treat a word that falls into both categories at once? Which rule takes precedence? No doubt there are principles of linguistics that would settle this issue, but it is easier simply to find an example. How do people actually pronounce a *war* word that is also an *ark* word? Wrack your brain for examples! I tried for months, without success. Then one sunny day, while raking grass cuttings, it suddenly came to me: *bulwark!*

A wave of joy passed over me. The issue could be settled empirically. But my joy was short-lived. "How," I introspected, "do you pronounce that word?"

"Rhymes with *pork*," I answered confidently.

"No," I reconsidered. "Rhymes with *park*."

Or could it rhyme with *perk* and therefore, God help us, with *quirk*? Could *quark* be a homophone of *quirk*? Can you imagine the announcement coming triumphantly out of Geneva that they had, at long last, discovered the "top quirk"?

Not being a nautical man, the fact is I may never in my life have uttered the word *bulwark* or heard it spoken. To be sure, it is used metaphorically with some regularity, but more often than not you come upon it only in writing. Dictionaries proved to be of no help, offering a range of acceptable vowel sounds for *bulwark*, given in an ordering that no two agreed upon. My research seemed to have reached a dead end.

Then, just a few weeks ago, coming upon a group of graduate students engaged in unproductive pursuits, I attempted to jar them out of their idleness with a challenge to find a *war* word that was also an *ark* word. Within 24 hours I was presented with *Newark*. (I digress to remark that one of the joys of Cornell is that the graduate students do not let you down.) Could the fate of the standard model then indeed hinge on the existence of a top quirk? No, I realized almost at once, in *Newark* the *ark* is unaccented, which disqualifies it as an example. Proudly and professorially I trundled out *bulwark* to illustrate what they should have been looking for, only to realize that it suffered from exactly the same defect.

To this day I have found no *ark* word that is also a *war* word with the accent on the *ar*. I appeal to you, dear reader, to wrack your brains for a specimen. Until one turns up I am working from the hypothesis that not only are quarks the fundamental building blocks of hadrons, but, perhaps even more remarkably, *quark* is the only *war* word that is also an *ark* word. Have I therefore given up?

Hardly!

Consider the word *cart*. It rhymes with *art*. So do *chart, dart, fart, hart, mart, part, smart, start* and *tart*. But not *wart*. Better still, not *quart*! Thus if we replace *k* with *t* we find two examples of the fact that when there is a clash, the *war* rule is more powerful than the *art* rule.

"Big deal," you may say, if you belong to the *ark* school, "*k*" is not *t*."

"Well," say I, "try another case." Consider *farm*. It rhymes with *arm*. So do *harm, alarm* and (appropriately enough) *charm*. But not *warm* and not *swarm*. The *war* rule is more powerful than the *arm* rule.

The harder you work on it, the clearer the trend becomes: *bard, card, hard, lard, chard* and *shard*, but *ward; carp, sharp* and *tarp*, but *warp; barf*, but *wharf; barn*, but *warn*.

I rest my case. No rational person can deny that the data point overwhelmingly to:

The warx *rule*: Given any consonant *x*, the vowel in all accented *arx* syllables is pronounced the same *except* when a *w* sound (*w*, *wh* or *qu*) precedes such a syllable, in which case the vowel is pronounced as in *war*.

It happens (unless, dear reader, you can rise to my challenge) that when *x* = *k* the *warx* rule operates only through the single example of *quark*. But should particle physicists, of all people, dedicated in their bones to the extraction of order and pattern out of the subatomic chaos, violate their deeply held and altogether admirable principles in their linguistic practices? Of course not! In the United States of America, *quark*, I must insist, rhymes with *pork*, not *park*.

15. Postscript

1. *Quarrel* elicited many letters to the editor of *Physics Today* (September 1994) as well as personal mail and email. I was sent many spurious counterexamples, with the *ar* syllable either unaccented, or followed by a second r. I was also told of the villages of Warkentin, Warkton, and even simple Wark, all in lands where they do not speak standard American. Only one putative counterexample worried me: "Edwardian." I admit to always having pronounced its second syllable to rhyme with "card". All my English friends claim it does rhyme with "ward", but according to my own rules their opinions don't count. Most Americans never utter the term. So until somebody convinces me otherwise I've added it to the long list of words I've somehow managed to mispronounce all my life. Should this position become untenable, I shall fall back on the fact that proper names are notorious for defying general rules.

2. I received a gratifying letter from Geoffrey K. Pullum, a distinguished linguist, then at Santa Cruz, now at Edinburgh:

> One of the things that drives theoretical linguists nuts is that the study of language is so often taken to be a subject that any person who went to college, even for a semester or two, majoring in anything, is able to contribute to. Nobody imagines that their opinion about the behavior of liquid helium is of interest if they're not a physicist. It's as if nothing can get it across to people that if they're interested in language as an empirical phenomenon—formally patterned behavior that actually occurs in the world—then you need *evidence* and *argument* to back up the claims you make. So we keep finding laypeople—the ordinary cab drivers, stockbrokers, and physicists that we meet when we travel among the common people—who regard themselves as authorities on language and want to proffer their opinion.
>
> But the trouble with you is that you're actually good at it. You do provide evidence and argumentation. You would actually be a great student in our subject. Too bad physics snapped you up, actually. It was when you discovered rule interaction ('the *war* rule is more powerful than the *ark* rule') that I realized I was dealing with someone who could have been a phonological theorist.

I basked in this unexpected praise. But I got my comeuppance when I ran into the great Victor F. Weisskopf at a reception in Ithaca, several months after *Quarrel* had appeared. "I thought your last column in *Physics Today* was silly," he said, drawing out that last word into a verbal stiletto.

"Well," I stammered, "an eminent linguist wrote me to say that my methodology was remarkable for an amateur."

"Linguistics," replied Viki, "is also silly."

16

What's wrong with this temptation

Once upon a time everybody knew why measurements in quantum mechanics don't reveal pre-existing properties. It was because the act of acquiring knowledge unavoidably messes up the object being studied. What you learn is not intrinsic to the object, but a joint manifestation of the object and how you probe it to get your knowledge.

In 1935 this state of happy innocence was forever dispelled by Einstein, who with Boris Podolsky and Nathan Rosen discovered how to learn about an object by messing up only some stuff it left behind in a faraway place. They concluded that knowledge acquired in this way was indeed about pre-existing properties of the object, revealed—not created—by the act of probing the stuff left behind. Bohr, however, insisted their conclusion was unjustified, and 30 years later John Bell *proved* that no assignment of such pre-existing properties could agree with the quantitative predictions of quantum mechanics.

A couple of years ago Lucien Hardy [1] gave this tale an unexpected twist, by finding a charming variation of the Bell–EPR argument. Hardy's theorem is even simpler than the argument of Daniel Greenberger, Michael Home, and Anton Zeilinger that I enthused about in this column four years ago.[1] The reason he was able to pull the trick off, and the reason, I suspect, nobody had noticed so neat an argument for so long, is that Hardy's analysis applies to data that are not correlated strongly enough to support

---

1 Chapter 7.

the argument of EPR. But they do give rise to an argument every bit as seductive, which Hardy is then able to demolish with surprising ease. Parts of the formulation I give here of Hardy's *gedankenexperiment*[2] are similar to those of Henry Stapp [2] and Sheldon Goldstein [3].

We consider two particles that originate from a common source and fly apart to stations at the left and right ends of a long laboratory. At the left station we can experimentally determine the answer to one of two yes–no questions, A or B. There is a choice of two other yes–no questions, M or N, to be answered by experiment on the right. Hardy provides questions A, B, M, and N, and a two-particle state $|\Psi\rangle$ for which the answers to the questions have the following features:

(i) If the questions are B and N, the answers are sometimes both yes.

(ii) If the questions are either B and M or A and N, the answers are never both yes.

(iii) If the questions are A and M, the answers are never both no.

Though two correlated particles subject to local probes in two faraway places also appear in an EPR experiment, Hardy's experiment is interestingly different. In an EPR experiment correlations in the data make it possible to predict the answer to whichever question you ask at one end of the laboratory by asking a suitable question at the other end. In Hardy's experiment you cannot perform this trick. If, for example, you want to learn the answer to A without messing up the particle on the left, you can try getting the answer to M on the right. If that answer is no, then (iii) does indeed guarantee that the answer to A will be yes. But if the answer to M is yes, you cannot predict the answer to A. If you try instead to measure N on the right, you run up against the same

2 Thomas F. Jordan brought Hardy's work vividly to my attention in articles submitted to *Physical Review A* and the *American Journal of Physics*.

problem: You can predict with certainty the result of measuring
A on the left in only an unpredictable and uncontrollable fraction
of the runs. Similar difficulties arise if you try to learn the answer
to B from measurements on the right or the answer to M or to N
from measurements on the left. We have here what one might call
a semi-EPR situation.

But that semi-EPR situation leads one into temptation just as
irresistibly as the full-blown variety. The temptation emerges when
you imagine a series of runs in which one chooses the question at
each end of the laboratory by tossing a coin at that end after the
particles have left their common source but before they arrive at
the ends to be tested.

We all agree—even Bohr might agree—that something in the
common origin of the two particles must underlie the correlations described in (i)–(iii). Since the questions to be asked are
not picked until after the particles have left their source, the features of the particles responsible for those correlations cannot
depend in any given run on what happens when the coins are
tossed. Furthermore, since each question probes only one of the
particles, the answer to a question at one end of the laboratory
can be influenced only by features residing in the particle at that
end and not by features residing in the faraway particle at the
other end.

If you accept those last two sentences, then you are in trouble.
According to (i), in some of those runs in which the questions
end up being B and N, the answer to both is yes. In these particular runs the particle on the left is indisputably of a type that
allows the answer yes to question B, and the particle on the right
is of a type that allows the answer yes to question N. But in any of
these particular runs the tosses of the coins could have resulted
in questions B and M being asked instead. Since the particle on
the left is of a type that allows the answer yes to B, the particle
on the right must be of a variety that prohibits the answer yes to
M. Otherwise if the coins had come up differently, it would be

possible to get answers yes to both questions B and M, which (ii) forbids. By the same token, since the tosses of the coin could have resulted in A and N being asked, and the particle on the right allows the answer yes to N, the particle on the left must prohibit yes to A. But the tosses of the coin in any of these particular runs could also have resulted in questions A and M being asked. Since each particle in these runs is of a type that prohibits the answer yes to its question, the particles in such a run would have to give the answer no to both A and M. But that is precisely what (iii) forbids.

People who find Bell–EPR profoundly mysterious ought to find this state of affairs equally bizarre. Those immune to the charms of EPR will have stopped reading after my second paragraph. So since you, faithful reader, are eager to know what underlies this astonishing trick, let me tell you one way to do it.

Take A and M to be any nontrivial questions you like. Pick any four one-particle states that lie entirely in their yes and entirely in their no subspaces. Call them $|Ay\rangle$, $|An\rangle$, $|My\rangle$, and $|Mn\rangle$. Take the two-particle state $|\Psi\rangle$ to be a superposition of products of these yes and no eigenstates. We guarantee feature (iii) of the data by requiring the no–no state $|An, Mn\rangle$ to be absent from that superposition:

$$|\Psi\rangle = \alpha|Ay, Mn\rangle + \beta|An, My\rangle + \gamma|Ay, My\rangle. \qquad (1)$$

Take the question B to have a single yes eigenstate $|By\rangle$ that is a nontrivial linear combination of $|Ay\rangle$ and $|An\rangle$, and take N to have a single yes eigenstate $|Ny\rangle$ that is another such linear combination of $|My\rangle$ and $|Mn\rangle$. Feature (ii) requires $|\Psi\rangle$ to be orthogonal to $|By, My\rangle$,

$$0 = \langle By, My|\Psi\rangle = \beta\langle By|An\rangle + \gamma\langle By|Ay\rangle, \qquad (2)$$

and orthogonal to $|Ay, Ny\rangle$,

$$0 = \langle Ay, Ny|\Psi\rangle = \alpha\langle Ny|Mn\rangle + \gamma\langle Ny|My\rangle. \qquad (3)$$

Feature (i) requires $|\Psi\rangle$ *not* to be orthogonal to $|By, Ny\rangle$:

$$0 \neq p = |\langle By, Ny|\Psi\rangle|^2$$

$$= |\alpha\langle By|Ay\rangle\langle Ny|Mn\rangle + \beta\langle By|An\rangle\langle Ny|My\rangle + \gamma\langle By|Ay\rangle\langle Ny|My\rangle|^2, \quad (4)$$

which (2) and (3) reduce to

$$0 \neq p = |\gamma|^2 |\langle By|Ay\rangle|^2 |\langle Ny|My\rangle|^2. \quad (5)$$

This tells us that the coefficient γ must be nonzero, and (2) and (3) then tell us that β and α can't be zero either. So for any questions A and M we can perform Hardy's magic trick in any two-particle state of the form (1) with three nonzero coefficients.

How big can we make the probability p whose nonvanishing gets us into all this trouble? It follows directly from (2) and (3) and $|\alpha|^2 + |\beta|^2 + |\gamma|^2 = 1$ that

$$p = \frac{p_l(1-p_l)p_r(1-p_r)}{1-p_lp_r} \quad (6)$$

where $p_l = |\langle By|Ay\rangle|^2$ and $p_r = |\langle Ny|My\rangle|^2$. Maximizing (6) gives uniquely $p_l = p_r = (\sqrt{5}-1)/2 = 1/\tau$, where τ—would you believe it?—is the golden mean. This gives p, the fraction of BN runs in which both answers are yes, the maximum value $1/\tau^5 = 0.09017$. Nine percent of the time something happens that the correlations described in (ii) and (iii) would appear absolutely to prohibit. Sensational!

Some experts might question my enthusiasm, since constructing arguments like Bell's in the absence of perfect EPR correlations is old stuff, originally inspired by the inability of any real experiment to demonstrate that correlations are perfect. The semi-EPR argument in the Hardy state, such experts might maintain, is merely an example of a violation—and not a very strong one—of an inequality [4] that, though very plausible even in the absence of perfect correlations, has already been reported to be violated in many earlier experiments:

$$p(By, Ny) \leq p(By, My) + p(An, Mn) + p(Ay, Ny). \qquad (7)$$

To understand what it might mean to violate this inequality, imagine a world in which each particle carried information specifying its answer to either question it might be asked. Call a particle x if its answer for question X is yes. The small side of the inequality is the fraction of particles that are b and n. The first term on the large side is at least as big as the fraction that are b, n, and m, while the third is at least as big as the fraction that are b, n, and a. So we would have an upper bound if we were to add in the fraction that are b and n but neither a nor m. Since $p(An, Mn)$ is an upper bound for this last fraction, the inequality must hold in such a world. It is violated, however, in a Hardy state, with 0 on the right and a probability as big as 0.09 on the left.

An experiment to confirm such a violation in the optimal Hardy state will require detectors accurate enough to distinguish a 9% event rate from a rate of 0%. Earlier tests for violations in states inspired by EPR had an easier time of it, using questions that made the probability on the left $(2 + \sqrt{2})/8 = 43\%$ and the sum of probabilities on the right $3(2 - \sqrt{2})/8 = 22\%$. Hardy states will not lead to more definitive experiments.

But to rest with that conclusion is to fail to see what makes the Hardy experiment so charming. My quick explanation of the inequality (7) required each particle to carry information specifying its answers to two incompatible questions. This is not only grossly un-quantum mechanical but, in the absence of an EPR argument, not even especially plausible. Much thought has gone into relaxing the assumptions underlying this inequality, and it can be made much more compelling than I have bothered to do here. But the refutation of even those refined assumptions simply doesn't hit you with anything like the impact of Bell's old refutation of the EPR argument or Hardy's new semi-EPR demolition job. So although Hardy's questions provide a rather weak basis for a laboratory violation of the experimentally relevant inequality, they reign supreme in the *gedanken* realm. There they achieve their effectiveness not by refuting the subtle assumptions behind the inequality, but by leading you down the garden path every bit as enticingly as the full EPR argument does and then turning around and kicking you out of the garden with unprecedented efficiency and force.

References

1. L. Hardy, *Phys. Rev. Lett.* 68, 2981 (1992); 71, 1665 (1993).
2. H. Stapp, *Mind, Matter and Quantum Mechanics*, New York: Springer-Verlag (1993), p. 5.
3. S. Goldstein, *Phys. Rev. Lett.* 72, 1951 (1994).
4. J. F. Clauser, M. A. Horne, A. Shimony, R. A. Holt, *Phys. Rev. Lett.* 23, 880 (1969).

16. Postscript

1. This is a companion piece to Chapter 7, and, like *Elements of Reality*, it is more technical than usual for *Physics Today*.

2. When *Temptation* appeared in print, punctuation marks (periods) were missing from Equations (1) and (7).[3] I noted this in a letter to the editor in the November 1994 issue:

> Having argued forcefully in these pages[4] that one should refuse to publish in journals that do not properly punctuate their equations, I was startled to find the periods removed from the published versions of two sentence-ending equations in my June *Reference Frame* column. I would like to state for the record that all equations were properly punctuated in the manuscript. When the punctuation disappeared in the first set of proofs I asked that it be restored.

3 They have been corrected in this reprinting.
4 Chapter 6.

Physics Today agreed and provided me with a second set of proofs showing properly punctuated equations. What happened next is anybody's guess. Were I paranoid I would suspect malign forces ready to stop at nothing to advance the antipunctuationist agenda. Not being paranoid I view it as an unfortunate accident, like (but happily less momentous than) the famous 18-minute gap in Nixon's White House tapes.

17

What's wrong with this sustaining myth

I have a colleague who goes around declaring that the laws of physics require consciousness to cease with the death of the body. What he really means is that although he has no idea what underlies the phenomenon of consciousness, he can't imagine it's more than an extremely subtle manifestation of physiological processes that come to a halt when the body does. I'd be inclined to agree if he'd put it that way, but he doesn't. He insists on saying "Science has shown it," which I take to be shorthand for "Stop thinking and believe me." He invokes "science" as a blessing to sanctify what he says, or as a club to beat into submission those he disagrees with.

The public should be warned about such abuses of the name of science, and two sociologists, Harry Collins and Trevor Pinch, have set out to do that. "What everyone should know about science" is the subtitle of *The Golem*, their award-winning book of essays.[1] Written "for the general reader who wants to know how science really works and to know how much authority to grant to experts," it is a central text in a growing controversy between scientists and those who study science. Collins and Pinch take as their image for science the mythical golem, a "lumbering fool who knows neither his own strength nor the extent of his clumsiness and ignorance ... not an evil creature, but a little daft." Their aim is to explain "what actually happens" in science. Prepare, they enjoin the reader, "to learn to love the bumbling giant for what it is."

1 Cambridge University Press (1993).

This is a fine goal. Scientists who set themselves up as sorcerers are a menace to the public and to science itself. People ought to have a better idea of what science can and cannot do. Unfortunately, however, though there are many fascinating tales about science in *The Golem*, Collins and Pinch infer from these studies a seriously deficient picture of the scientific enterprise. Here are some typical conclusions:

1. "Scientists at the research front cannot settle their disagreements through better experimentation, more knowledge, more advanced theories, or clearer thinking."
2. "The truth about the natural world [is] what the powerful believe to be the truth about the natural world."
3. Scientists "have, of course, their special area of expertise, the physical world, but their knowledge is no more immaculate than that of economists, health policy makers, police officers, legal advocates, weather forecasters, travel agents, car mechanics, or plumbers."

One could, of course, interpret these conclusions as virtually self-evident. The first merely characterizes the research front, that boundary between known and unknown, where disagreement among scientists is the order of the day. When disagreements are settled we are no longer at the research front. The second is valid because if the powerful in science persisted in believing what was false about the natural world, they would soon cease to be powerful. The third is a warning about people who wave the wand of Science to waft away opinions they do not share, and it is an injunction to respect the knowledge of all experts within their spheres of competence. On the whole, however, Collins and Pinch have something different in mind: "Science works the way it does not because of any absolute constraint from Nature, but because we make our science the way that we do."

The Golem is full of such declarations. Their collective import is this: It is naive to think that the aim of our enterprise is to discover things about nature or to frame concepts that capture important features of the world. What we are actually doing is constructing a consensus among our fellow investigators. Ours is one of the most effective processes of consensus building ever achieved, and one of the aims of sociology is to learn why it works so well. You and I may think it works because, by a long and arduous process, scientists have become better and better at formulating questions that extract useful information from the natural world while avoiding questions that lead nowhere. This view is an expression of our naive realism, but it is important that we believe it. The conviction that we are trying to learn an objective truth is a powerful sustaining myth that drives us onward in our efforts at consensus building.

If this sounds absurd to you, consider: Scientists do, in fact, build consensus out of disagreement by a social process. How could it be otherwise? Consensus is a social phenomenon. The notion that this is the whole story—that all we are doing is exercising our exceptional skills at coming to agreement—is a sustaining myth for sociologists. It leads them to reject facile explanations of how scientific controversies come to an end, and to examine more thoroughly the actual process by which we come to agree that "the truth of the matter" has been established.

The pertinent issue in assessing the claims of *The Golem* is not whether scientific truth is determined by constraints from nature or by social construction, but whether Collins and Pinch strike a satisfactory balance between these two aspects of the process. I believe their book furnishes an instructive demonstration of what can go wrong if you focus too strongly on the social perspective. By paying insufficient attention to how nature does constrain us, Collins and Pinch draw lessons about the building of scientific consensus that leave out an essential part of the story.

The authors of *The Golem* know, as do you and I, that much of what one reads about "scientific method" bears little relation to what actually happens. We rarely proceed by framing a hypothesis and devising an experiment to test it, rejecting the hypothesis if the test is not passed. This is a cartoon version of what we do. Our understanding may be too uncertain to frame a clear-cut test, our interpretation of experiments depends strongly on the conceptual context in which they were designed, and our data are often ambiguous and susceptible to a wide range of explanations, many of which have little or nothing to do with what the experiment was intended to probe.

Collins and Pinch illustrate this with case studies from a variety of disciplines. They typically follow a chain of difficult experiments intended to address a certain constellation of questions, noting the successes, the failures, the ambiguities, the disputes, going on for months, years, even decades. They find that even while the struggle rages, while doubts are still unresolved, a broad consensus can emerge about the issues that originally gave rise to those experiments. "It is always thus," they conclude, "science works by producing agreement among experts … Experiments in real science hardly ever produce a clear-cut conclusion—that is what we have shown … The mess [is] not allowed to be the message. At the end triumphalism rule[s]." This triumph of triumphalism is not confined to textbook oversimplifications, Nobel prize citations, or newspaper interviews. The claim of *The Golem* is that it lies at the very heart of the scientific enterprise.

Every case study in *The Golem* supports this picture of experimental studies plagued by conflict and ambiguity, and I'm sure most of us could write comparable case studies based on our own experience. Why indeed does the scientific community nevertheless often reach firm agreement on a question, long before a difficult experiment designed to explore it has come close to a definitive conclusion? Are we really just experts at negotiating myths? Or could something else be going on?

Read these essays. I think you will find the answer to be obvious (though I doubt it will be to the lay reader). Agreement is reached not just because scientists are so very good at agreeing to agree.

It is reached because many other things have been going on that Collins and Pinch have said nothing about—things outside the scope of their study. Even first-class studies of episodes in the history of science can't cover all parallel activities at the same level of scholarly detail. Furthermore, if attention were redirected toward a different set of related questions and experiments, those too might well provide another case study supporting the same view of how science operates. To a first approximation, it is "always thus" because even though many clues in a complex network of evidence will always be far from definitive, the probability of a conclusion supported by a multitude of interlocking mutually reinforcing clues can still be close to certainty.

I don't have to elaborate on this for readers of *Physics Today*, but the point ought to be made more extensively to "the general reader who wants to know how science really works." The method of Collins and Pinch is to follow one strand of an enormous tapestry of fact and analysis. They note that the strand is quite thin in places. Often they can demonstrate that the contribution of that particular thread to the whole picture has been greatly exaggerated. But they pay only perfunctory attention to everything else that holds the tapestry together. They never acknowledge that an enormous multiplicity of strands of evidence, many of them weak and ambiguous, can make a coherent logical bond whose strength is enormous. On the few occasions when they hint at this, the resulting consensus is nevertheless attributed not to reason, but to internal politics.

Collins and Pinch are smart people and they have some fascinating stories that they tell very well. They say they love science, they know a lot about it, and they may be starting to have a serious effect on how people think about it. But their own sustaining myth of the social construction of scientific truth has lulled them into finishing their case studies with an incomplete story of how science acquires knowledge about the world. By focusing exclusively on individual threads, they have produced a picture of "what actually happens in science" that overlooks the crucial role played by the intricate structure of the whole interconnected tapestry.

The view of science that emerges from this oversight is succinctly expressed in the fable with which Collins and Pinch conclude *The Golem*. A class of children—"a microcosm of frontier science"—all try to measure the boiling point of water. In the last ten minutes of the lesson, the teacher collects their disparate results and, without performing the experiment herself, persuades the children that "their experiment has proved that water boils at exactly 100° C." And, Collins and Pinch explain to their general reader, "that ten minutes illustrates better the tricks of professional frontier science than any university or commercial laboratory ... Eddington, Michelson ... are [the schoolchildren] with clean white coats and 'PhD' after their names ... There are theorists hovering around, like the schoolteacher, to explain and try to reconcile. In the end, however, it is the scientific community (the head teacher?) who brings order to this chaos, transmuting the clumsy antics of the collective Golem Science into a neat and tidy scientific myth."

That, the general reader is told at the end of *The Golem*, is "most of what there is to know about the sociology of science."

In next month's *Reference Frame*, I'll examine how *The Golem* treats a subject we all know something about, the theory of relativity, to illustrate how Collins and Pinch go about constructing such a case.

17. Postscript

1. With this column and the next I stumbled into the dispute between scientists and the scholars who study science that was to be called "the science wars." I had not seen the notorious book by Paul Gross and Norman Levitt[2] that had come out a few

---

2 *Higher Superstition: The Academic Left and Its Quarrels With Science*, Johns Hopkins University Press (1988).

years earlier. Alan Sokal's famous hoax, inspired by his reading of Gross and Levitt, hit the front page of the *New York Times* a month or two after my columns appeared in *Physics Today*.

2. Looking through those files of mine that have survived two decades of computer upgrades, I find that the views of Collins and Pinch couldn't have come as a complete surprise to me. In the summer of 1993 I had participated in a week-long series of all-day seminars in which a group of faculty from Science and Technology Studies discussed their views with Cornell scientists from many different disciplines. At the end of the week I submitted a report:

What I learned this week about science

Scientists are a group of misogynist racists suffering from the collective delusion that there is a world external to themselves which exhibits uniformities that they can describe and understand. To perform this task they raise money from warmongers and exploit the labors of people whose existence they publicly deny. Skilled in the arts of cheating and lying to each other, they are equally adept at deceiving the general public, and have become proficient at misadvising legislatures and the courts. When unable to come to agreement they engage in polemics of exceptional ferocity until only one of the contending views survives. This may account for their otherwise mysterious ability to arrive at a consensus on how to make bridges that fall down, airplanes that crash, chemicals that poison and pollute, ideologies that preserve the existing order, and bombs that obliterate it.

18

The golemization of relativity

"Two Experiments that 'Proved' Relativity" is Chapter 2 of *The Golem: What Everyone should Know about Science*. This prize-winning book of essays by the sociologists Harry Collins and Trevor Pinch[1] is at the center of an ongoing debate between scientists and those who study science. In Chapter 2, Collins and Pinch use examples from the history of relativity to show the lay reader how they believe science reaches its conclusions. Here I use their essay to show readers of *Physics Today* how I believe Collins and Pinch reach theirs.

The two experiments are the Michelson–Morley experiment and the Eddington solar eclipse expedition of 1919. I learned a lot about both from Collins and Pinch, but I found unconvincing the lessons about the nature of science they draw from these studies of relativity. I focus on their treatment of the Michelson–Morley experiment, and how it bears on the acceptance of special relativity by the scientific community.

The presentation of relativity in *The Golem* starts with a peculiar statement of the principle of the constancy of the velocity of light: "Einstein's insight [was] that light must travel at the same speed in all directions."

This is the kind of simplification anyone might make in presenting a technical matter to the lay reader. But this particular reformulation of the postulate that the speed of light is independent of the

---

1 Cambridge University Press (1993). I address some of the general issues raised by *The Golem* in Chapter 17.

speed of the source also happens to reinforce the view that Collins and Pinch develop—that doubts about the Michelson–Morley result put at risk the logical foundations of relativity. The lay reader is told little about the actual content of special relativity. Time dilation and length contraction are briefly mentioned. The inclination everybody has on a first hearing—to regard these phenomena as outlandish—is heightened by qualifications such as "if Einstein's ideas are correct" and "if the theory is correct." To this touch of skepticism Collins and Pinch add a dash of moral indignation by also mentioning the "sinister" mass–energy relation and the explosion of the atomic bomb. The bomb is cited as an incontrovertible piece of evidence for the validity of relativity. No other compelling evidence is offered. Nothing is said, for example, about the unity and coherence that relativity immediately brought to electrodynamics, about early confirmation of the relativistic forms for the energy and momentum of high-speed electrons, about the importance of relativistic corrections in the theory of atomic spectra, about lengthened lifetimes of rapidly moving unstable particles, about precise pre-Hiroshima confirmation of the mass-energy relation in nuclear physics or about the necessity for relativistic mechanics in the design of particle accelerators.

Collins and Pinch introduce the Michelson–Morley experiment with "the way the story is most often told… ." We are not informed who is telling the story or to whom. I imagine lay readers will take this to refer to how scientists view the development of relativity, rather than merely (but correctly) to describe how the subject is often presented in popular writings and texts.

The essay is haunted by such unspecified speakers and audiences: "The drama [of Michelson–Morley and the Eddington expedition] lies in the clarity and decisiveness of the questions and the answers." Drama for whom? For newspaper readers or for physicists? I presume the former, but to reach the point Collins and Pinch are heading for it is crucial for it to be the latter. Such failures to distinguish between the views of the scientists and the views of the public persist throughout the essay: "But even these experiments turned out to be far less decisive than is generally believed." Believed by newspaper readers? Or believed by physicists?

Collins and Pinch argue that the Michelson–Morley experiment played a crucial early role in establishing the validity of relativity, but that actually the experiment was extremely difficult to do, the results were ambiguous, and attempts to repeat it over the next 70 years produced a mixed bag of results. The fact that relativity was nevertheless widely accepted by the early 1930s shows, they maintain, that the establishment of scientific truth is a cultural process rather than a consequence of the accumulation of facts about the natural world. In support of this, we are given a long, fascinating history of Michelson–Morley and the many attempts to replicate and refine the experiment, with varying degrees of success, continuing all the way into the 1960s.

But although many texts and popular books do indeed treat the Michelson–Morley experiment as a fundamental pillar on which special relativity rests—the rock on which Einstein founded his church—it surely played no more than a minor role in why special relativity had become an absolutely fundamental part of physics by 1933. Collins and Pinch are aware of the marginal role of Michelson–Morley in Einstein's own thinking, but they do hold it to have been decisive for "others": "Although the experiment is often thought of as giving rise to a problem that Einstein set out to solve, this too is probably false. The link between Einstein and Michelson was forged by others some twenty or more years after the first 'decisive' experiments were completed."

We are not told who those "others" were, nor are we told who was responsible for the fact that "Only after Einstein's famous papers ... did the experiment become 'retrospectively reconstructed' as a famous and decisive proof of relativity." The general reader is likely to conclude that these retrospective reconstructors must also have been physicists at the research front. If, on the contrary (and as seems to me more likely), they were journalists and textbook authors, then where does this leave the case for the crucial importance of Michelson–Morley?

Much is made of a 1925 rerun of Michelson–Morley by Dayton Miller that "was widely hailed as, effectively, disproving relativity." "Widely hailed" means "was awarded a prize by the AAAS," but the general reader is not helped to assess the significance or context of the prize. (I looked it up—it was for the best talk given at the annual meeting.) The existence of a variety of reasons for believing in special relativity by 1925, having nothing to do with Michelson–Morley, is simply not mentioned.

In 1933 Miller published an extensive article in *Reviews of Modern Physics* casting doubt on the old null result, but

the argument in physics was over by 1933. Other tests of relativity, including the Eddington observations of 1919, indirectly bolstered the idea that the theory of relativity was correct and that the velocity of light must be constant in all directions. The sheer momentum of the new way in which physics was done—the culture of life in the physics community—meant that Miller's experimental results were irrelevant.

Here the reader is finally told there have been other tests of special relativity. The only one actually mentioned, however, is the Eddington expedition of 1919, which had nothing to do with special relativity, and whose results, in any event, are persuasively argued in the second half of the essay to be less than conclusive. The word "indirectly" might refer to this weak link, or it might be meant to characterize the considerable array of "other tests" available by 1933. This substantial body of evidence is not only unmentioned but can be dismissed as "indirect" compared with Michelson–Morley by virtue of the translation of "Einstein's insight" into the direction independence of the velocity of light. Relativity, the reader is left to believe, had somehow just got into the culture of the physics community, so by 1933 poor Miller never had a chance.

A moral that Collins and Pinch draw from their story is that "The meaning of an experimental result does not, then, depend only upon the care with which it is designed and carried out, it depends upon what people are ready to believe."

Yes indeed. But what brings people into a state of readiness to believe or disbelieve an experimental result and its interpretation? The lay reader can only guess. Hardly a hint is given that this essay follows a single tiny strand in an enormous tapestry of fact and analysis. Collins and Pinch correctly note that the importance of the strand to the whole picture has been greatly exaggerated, and that the strand is quite thin in many places. If they stopped there, I would have no disagreement with them. But they are quite emphatic about the fact that relativity was accepted in spite of the weakness of this strand, and without even a glimpse of the rest of the tapestry, I do not see how the lay reader can fail to conclude that relativity is fraudulent.

To be sure, Collins and Pinch state at the end that "we have no reason to think that relativity is anything but the truth and a very beautiful, delightful and astonishing truth it is ... ." On the basis of what they have said, however, one can only wonder why they might think so. The sole reason for believing in relativity that they cite as compelling is the atomic bomb, neither beautiful nor delightful, as indirect as evidence can be, and unavailable in 1933 when the "argument in physics" was already over. It is only by presenting legitimate doubts about the Michelson–Morley experiment without any serious discussion of other reasons for believing in the special theory that Collins and Pinch are able to make their case that the acceptance of the theory was primarily a cultural phenomenon, "just as much a license for observing the world in a certain way as a consequence of those observations ... Thus was the culture of science changed into what we now count as the truth about space, time, and gravity."

One of the tantalizing things about this essay for the physicist is that Collins and Pinch implicitly raise some interesting questions about the history and practice of relativity. How indeed did Michelson–Morley assume the special status that many texts and popular books assign to it? How does this fit into the more general tendency of texts and popular books to create mythologies by pedagogical shortcuts or oversimplifications for the general reader? Do such mythologies have an impact on the work of practicing scientists? Instead of addressing these questions Collins and Pinch confuse a legend about relativity with relativity itself, and they use the weakness of that legend to undermine the intellectual coherence of the process by which the theory was established, paving the way for them to transform it from a truth about the natural world into a cultural artifact.

18. Postscript

1. The two columns on *The Golem* resulted in many letters to the editor in July 1996 and January 1997. These included responses from the authors, Harry Collins and Trevor Pinch, and further comments from me.

2. A few years later, I summarized how I myself would have put the point Collins and Pinch make about the acceptance of relativity and the reaction to Dayton Miller's 1933 experiment:

> In 1905 Einstein put forth the special theory of relativity, which recognized that some problems in the foundations of electrodynamics could be solved by a simple but radical refinement of the concept of time. Relativity requires the Michelson–Morley experiment to have a null result, which it did to within the precision of the experiment. Because the Michelson–Morley experiment can be simply related to one of the two postulates Einstein used in his original formulation of the theory, a pedagogical tradition developed of presenting the theory as a deduction from that null result. In its first two decades the special theory of relativity became an accepted part of physics because of its success in restoring coherence to electrodynamics, because of the observation of the gross deviations

from Newtonian mechanics that it required in the behavior of
rapidly moving electrons, because of its successful application to
the new quantum theory of atomic spectra, and because of the
qualitative and then quantitative observation of deviations of
the masses of atomic nuclei from the sums of the masses of their
constituent parts. Thus when Dayton Miller published a paper
in 1933 claiming a small nonzero effect in a new version of the
Michelson–Morley experiment, it made far more sense to assume
some source of error in his very difficult experiment, than to
abandon a theory that had been so broadly successful.

3. In the two decades since I wrote Chapters 17 and 18, in my
   efforts to make sense of quantum mechanics, I've come to take
   a more complex view of the role played by scientists themselves
   in the nature of their science. See Chapters 30–33.

19

Diary of a Nobel guest

Friday, December 6th. Arrive Stockholm 8 am, luggage stuffed with white tie costume, dark suits, evening gowns, newly acquired white shirts and ties. Light rain. Grand Hotel selectively grand. Bathroom magnificent but closets insufficient for two Nobel guests. No bureaus whatever. After dinner find hosts and Cornell physics colleagues Dave Lee and Bob Richardson newly arrived from Göteborg, wearing tiny gold lapel pins so reporters, autograph collectors can tell laureates from guests. Bob has bad cold. Get perfect 8 hours sleep, but first night always easy.

Saturday, December 7th. Breakfast buffet at Grand phenomenal, and attended by many old friends from glory days of superfluid helium-3. Black stretch limos—one per laureate—take physics and chemistry winners to lectures. Guests follow in tour buses. Lecture hall surprisingly small. Front rows reserved for Nobel guests. Physics lectures evocative of scientific memories from early 1970s. Bliss was it in that dawn to be alive. Chemistry talks also fun since buckyballs really physics. Or is superfluid ^{3}He really chemistry? Both prizes for something discovered accidentally while looking for something else. Back to Grand in dark. Get report on literature prize lecture by Wislawa Szymborska from those who cut chemistry to attend. Who would have expected parallel sessions? Awake half the night.

Sunday, December 8th. Laureates busy all day; guests free, weather dry, city beautiful. Collapse at 2 pm, awakening from nap in darkness at 3. Bus to opulent reception. Reunion of old ^{3}He crowd at delicious dinner. Laureates can't make it, having mandatory "Informal dinner (dark business suit)" at Academy of Sciences. Awake most of night.

Monday, December 9th. Guest status good for front row seats at economics lecture. Theory of auctions. Integrals and derivatives. Like physics except physics works. American laureates have lunch with ambassador. Poor laureates. Guests have learned to skip lunch between breakfast buffet and late afternoon reception. Today's event dwarfs yesterday's: Apotheosis of Informal Dress. Black suit blends right in. Succumb to earthly delights until time to depart for dinner in gorgeous baroque clubroom with Swedish Cornell alumni. Sleep all night.

Tuesday, December 10th. Big day. Women have hair set in morning and are confined to quarters until afternoon. Laureates off at a mandatory rehearsal (casual). Take long walk along water to check out City Hall. Many delivery vans, mysterious stacks of wood. Strange waterfront sculptures. Return to Grand via Café Access. Only bargain in Sweden: 20 kronor for half hour on internet (bratwurst from street peddler costs 25). Treatment shifts from brusque to cordial when waiter learns he is logging in Nobel guest. Send email greetings to kids, bulletin of events to Cornell Physics Department. Back at Grand don required costumes and off to Concert Hall at 3, laureates in limos, guests in "Nobelbil" (old buses with new signs). As guests of oldest laureate in number one field, get front row center balcony seats. Most opulent production of *Die Fledermaus* falls far short of tableau. Concert Hall glows. Members of Swedish Academy seated on stage. Full orchestra and singers ready in great balcony above it. Jewels and medals in all directions. Check abundant flowers festooning edges of balcony. Real, of course. Is this a dream?

Flourish of trumpets promptly at 4 followed by entrance of laureates to Mozart march, led by Cornell delegation. Burst unexpectedly into tears of joy. Remarkably like wedding. Royal music follows. All rise for King, Queen, and Royal Aunt. Swedish heard for first time in festivities, but pamphlets with English text provided. Proudly recognize sentence of mine from old nominating letter in physics section of English libretto. Each laureate engages in complex hand maneuvers with King. Attendant hands King large citation with box containing medal on top. King extends same to laureate with left hand. Laureate grasps other side of citation with own left hand. Right hands make direct contact in shake beneath citation. After shake King lets go, laureate holds citation. Medal does not fall off. Trumpet flourishes while laureate bows to King, members of Academy, and audience. Warm applause as laureate returns to seat. After physics, chemistry come Sibelius, medicine, Grieg, literature, Mozart, economics. Szymborska bows in wrong sequence, thereby earning prolonged ovation. Don Giovanni tops it all off by singing seductively from on high accompanied by Leporello playing on real mandolin. All rise for National Anthem as King, Queen, and Royal Aunt leave. Photographers rush onto stage. Much shaking of hands. Evokes post-debate ritual in recent American presidential election.

Make way past throngs of Stockholmers to great line of red municipal buses, each become Nobelbil for occasion. Student riders kindly offer seats to ladies. Glance at fellow white-tie straphangers. Was ever stranger sight seen in Arctic night? Off to City Hall now transformed into enchanted palace. Logs have become great bonfires on waterfront. Sinuous pieces of sculpture noted by water in morning now ablaze. From courtyard entrance to top of great stairway in opposite corner is honor guard of candle-carrying Cub Scouts. Smile at scouts, frown at scouts, salute scouts. No eye contact. Serious business.

Dinner guests (1300) easily seated with aid of 71-page book of alphabetic lists and maps of hall and all 66 tables. Orchestra launches into *Marche triomphale* and diners rise, as from high up in far corner of roofed-in Venetian courtyard slow procession emerges onto balcony, traverses length of side wall, and descends monumental staircase at far wall, to take places at head table stretching back nearly full length of floor. Music ends, King sits, so do all. "Do you mind if I smoke?" asks woman on left and realize am coming down with sore throat. "No, of course not," reply gallantly, and she chain smokes her way through rest of evening. After toast to King ("To the King!") and toast from His Majesty to Alfred Nobel ("Alfred Nobel!"), wild rumpus starts. Each course begins with musical procession of 150 waiters following same lofty course taken by occupants of head table, but fanning out to cover entire floor. Exercise carried out under direction of wait-conductor positioned high on stairway. Effect uncanny. All across vast hall clatters of serving spoons against platter start and stop simultaneously. Descent of final course (*glace Nobel*) astonishing. Lights dim. Two great canopies flutter out of nowhere to form roof within roof over great staircase. Staircase engulfed in cascade of waist-deep white cloud pouring down from top, heralding appearance of gentlemen in Turkish garb leading proud and enormous dogs. Their stately descent followed by costumed singers and entourage who give us exquisite solos, spine-tingling duet, and dazzling quartet. As sopranos ascend on bench to point halfway between floor and ceiling, still courageously belting it out, there appear at top of still smoking stairs not elephants (if ceremony in Concert Hall was *Die Fledermaus*, then banquet is *Aida*) but 150 waiters bearing massive glowing trays of ice cream for revelers below Waiters descend, singers ascend, spoons clink, plates clatter, and a woman's voice whispers in my ear, "You're the last one—would you like everything left on my tray?" Gasp "Yes" gratefully, and soothe aching throat with triple portion of *panache de sorbet aux mûres sauvages des champs et de parfait à la vanille* as my companion lights up yet another.

Brief after-dinner "two minute" speeches by eldest laureate in each category are anticlimax. Only Szymborska honors time limit. Meditate on narrow line between magnificence and bad taste. This unquestionably magnificent, but requires monarchy to pull off.

King rises as do all, and occupants of head table exit over reverse of entrance route. Dancing follows atop great staircase in enormous golden hall. Laureates back at work, granting TV interviews in back rooms. Exhausted, catch 11 o'clock bus back to Grand, resisting temptation to move on to "Nobel Nightcap" post-party student party. Doffing finery, am Cinderella again, and lie awake all night, with visions of sugarplums and bad cough.

Wednesday, December 11th. Stagger off to breakfast buffet. Laureates there, bright and chipper after Nobel Nightcap. Some left at 3 am. Bravest stayed until 5. Realize if economics laureate William Vickrey had not died within days of hearing of prize, award ceremony would have finished him anyway. After breakfast catch bus for luncheon buffet. "Terminal dyspepsia," mutters companion. Passing through Grand in evening, encounter astounding, indestructible laureates, smiling and spiffy in white tie and tails again, ready for dinner with King at palace. Stagger off to own dinner. Up all night coughing and sneezing. Have acquired full-blown Nobel cold.

Thursday, December 12th. Realize must spend day in bed. Sleep noon to 2. Awake refreshed. Long walk, museum, long walk back, dinner, bed. Awake all night. Shouldn't take naps.

Friday, December 13th. Santa Lucia Day. Pass ascending procession of candle-carrying maidens in white gowns en route to breakfast. Lead maiden wears five-candle crown (centered square). Accompanying representative of Nobel Foundation explains early morning visits in bedroom mandatory for laureates, optional for guests, if ordered night before. (From concierge? Room service? Housekeeping?) Maidens appear later at breakfast. Lights dim. Distant voices intone famous Neapolitan song at half tempo. Maidens enter and provide 10-minute serenade of *a cappella* Renaissance tunes and Christmas ditties in candlelight. Final number reprise of entering *Santa Lucia*, in course of which five-candled maiden slowly turns and leads crew out. As final distant *Santa Lucia* fades to whisper, lights go back on. Wolf down morning repast of oatmeal with honey (for sore throat), herring, prosciutto, assorted cheeses, fruits, pastries, in time to catch bus for Uppsala. Temperature well below freezing. Snow starts. Waive optional tour of frigid Uppsala for warmth of auditorium where physics laureates give miniversions of Nobel lectures. At post-lecture reception Richardson recommends strong drink for laryngitis. Cured his. Diverse fluids flow at sumptuous luncheon in great hall of Castle, but laryngitis gets worse. Not strong enough. Doze on bus back to Stockholm. Gets colder and colder.

Back at Grand, handful of remaining guests set out in full formal regalia for Lucia Dinner given by students at Stockholm University Union. Can do nothing but croak, not entirely inappropriately, since pervading theme of evening is Smiling and Jumping Little Green Frog. Confusion reigns at banquet, scaled-down version of Tuesday's extravaganza. Meal punctuated by endless series of toasts, each accompanied by song, lustily belted out by toast-masters, diners, and guests who follow with Swedish librettos. Throat aching and mindful of advice, take double dose of schnapps at each of first three toasts. Assuredly beneficial effects on larynx are countered by increased desire to join in singing, now accompanied by linking of arms, and rhythmic longitudinal swaying. Wine and beer ease pain again, ice cream appears in great burst of sparks, lights go out and soft, slow distant sounds of *Santa Lucia* float down stairs, followed by different five-candled maiden and white-gowned retinue. Litany of sweet and mournful tunes unexpectedly modulates into round of finger-snapping, white-gowned, hip-wiggling hot cha-cha, which just as abruptly flips back into final worshipful *Santa Lucia* as maidens slowly and softly float away back up stairs.

More toasts follow. Swaying now transverse as well as longitudinal. If Tuesday's banquet *Aida*, then this one *Faust*. Toastmaster announces bus that was to meet us at midnight will not appear until 1. Good thing. Between midnight and 1 culminating event of week takes place: induction of laureates into Order of Ever Smiling and Jumping Little Green Frog. Bedlam of scene impossible to convey. Culminates in all six 1996 Nobel Laureates in physics and chemistry lined up together, uttering cries of "Ribbit, Ribbit," while squatting on haunches and jumping up and down. Sensational ending to can-you-top-this week.

Saturday, December 14th. Go out into freezing night at 1 am. No bus. Toastmaster has lied. No taxis. Peer into frosty darkness. Nothing. Try to thumb rides as six black limos pull out, but all are full. Representative of Nobel Foundation appears, waves magic cellular phone, and unscheduled municipal bus pulls up at Student Union and takes us home. Get to bed at 3. Sleep like log! Dream am King. Arise at 7. Sore throat gone! Take van to airport. Croak to SAS representative that it will do me no good to be seated two rows from smoking section. Sorry, she says, plane full, but hold on. Get bumped up to business class and fly back amidst parting salvos of salmon, herring, and champagne.

19. Postscript

1. In the early 1970s two of my Cornell Physics Department colleagues, Dave Lee and Bob Richardson, with their graduate student Doug Osheroff, made a discovery (superfluidity in helium-3) which clearly met all the criteria for a Nobel Prize. It took a quarter of a century for the physics Nobel Prize Committee finally to acknowledge this, in 1996.

Shortly after the announcement in October, the three prize-winners invited several of us who had worked with them in the 70s on aspects of their discovery to be their guests at the festivities in Stockholm in early December.

When I received a call from Richardson inviting me and my wife, I worried that we already had a crowd of Thanksgiving guests visiting us in late November, and a trip with a Costa Rican friend to his homeland irrevocably set for early January. I reported Bob's call to Dorothy as a vexing complication at a busy time. "Should we accept?" I asked. "Are you kidding!" she replied. So we went.

I suffered from jet lag for almost our entire week in Stockholm, and entertained myself during sleepless nights by writing down the events of the preceding day. At the end of the week I had enough for a pair of *Reference Frame* columns, but *Physics Today* decided one was enough. My original diary can be found below as Chapter 36.

2. Shortly after *Diary* appeared, I received an email from a member of the Swedish Academy. He was pleased by my characterization of the difference between physics and economics. He had disapproved of the questionable establishment of an Economics Prize in 1969, even though it had officially been called the "Nobel *Memorial* Prize." This attempt to distinguish it from the real thing proved unsuccessful.

3. One of the other Cornell guests at the festivities, who knows his breeds better than I do, complained that I underestimated their shock value by failing to identify the "proud and enormous dogs" at the banquet as Afghan hounds.

4. A few years after the column I got a phone call in mid-October from a newly-minted Nobel Prize winner, whom I didn't know. He had read my column and wanted advice on what clothes to bring to Stockholm.

20

What's wrong with this reading

Many scientists pride themselves on their ability to read diffi-cult texts in areas of their discipline conceptually or histor-ically remote from their own. It has become evident in the recent "science wars" between scientists and science critics that this abil-ity can diminish rapidly with interdisciplinary distance. (I use the word "critics" in the neutral sense—cf. "theater critics." I can think of no other term encompassing the full array of practitioners from sociology, anthropology, history, literature, and cultural studies who have turned their attention to the activity of scientists.) Fronts are opening in the science wars on which some scientists are mis-representing and oversimplifying as egregiously as those at whom they direct their fire.

I shall illustrate this with one of the strangest and most notori-ous texts on the battlefield, Bruno Latour's "A Relativistic Account of Einstein's Relativity" [1]. This essay has been criticized by physicists for misconstruing the content of relativity and being filled with elementary technical mistakes. It is on display in Alan Sokal's famous spoof [2], one of its "mistakes" showed up in Steven Weinberg's much-cited article in the *New York Review* [3], and I know of two articles on the "Relativistic Account" scheduled to appear in anthologies devoted to the new and gloomy art of extracting technical errors from the writings of science critics.

I believe such attacks miss the point of Latour's essay. While I have not myself succeeded in making complete sense of it, there are texts by Nietzsche, Hegel, and Kant in which there is virtually nothing I can make sense of. Nevertheless, I have not concluded that they are charlatans. Critics of the science critics ought to exercise similar caution. The straightforward explicit style toward which scientists strive (and pick up any issue of *Science* to remind yourself how successful *we* are in achieving it) is inappropriate in disciplines where the objects and aims of inquiry have themselves an ambiguous and uncertain character.

Latour takes an anthropological slant on things. Physicists recently discovering his "relativistic account" are not the only ones he puzzles. Many distinguished British critics of science find him a far-from-easy read, and they have fired more accurate salvos in his direction than some of the interdisciplinary ballistic missiles I have seen launched from the science side.

His text focuses on a little book, *Relativity: The Special and General Theory*, written by Einstein in 1916 for the general reader. If I had been Latour's referee, I would have insisted (among other things) that he change his title to something like "What Can We Learn from a Popular Text by Einstein about the Study of Society?" I imagine Latour would have refused to make the change, because his title is much more fun.

Which brings us to the matter of fun. Bruno Latour is clearly a man who enjoys having fun. His article is always playful: "Although [Einstein] takes the reader, at the beginning, to Trafalgar Square, he is not interested in sending him to tail Hercule Poirot on to the train at Paddington …" This sets the tone and establishes the almost jocular but not necessarily inaccurate idiom in which somewhat more technical matters are put:

Playing the idiot, the author-in-the-text redefines what an event is …"

The only thing required of them is to watch the hands of their clocks closely and obstinately …

[Einstein's] panic at the idea that observers sent away might betray, might retain privileges …

[The] hard and lowly work of building a rigid scaffolding to frame an event …

What is a funny article doing in *Social Studies of Science*? I have tried from time to time to publish funny articles in *Physical Review Letters*. Only minuscule bits of the fun have escaped the editorial mangle. But different fields have different conventions. *Social Studies of Science* allows authors to be entertaining. When physicist critics of Latour fail to notice when he is being funny, they put at risk our proud reputation for having a finely tuned sense of humor.

Now on to the hard and lowly work of suggesting what Latour may actually be writing about. I defer here to my uniquely qualified daughter Liz, who has been in cultural studies for some years, is now in anthropology, and once taught a class at Harvard on relativity for nonscientists in which Einstein's little book served as a text. I am taking the outrageous paternal liberty of displaying below my edited version of her instant reading of Latour:

The big point from the social science perspective is the role of
the observer—the guy who is needed to know that the man on
the tracks and the man on the train don't say the same thing, and
who is in a position to compare their readings without saying that
one or the other is wrong. This is analogous to the social scientist
looking at society. What's interesting for the social scientists is that
it doesn't really matter how this observer is positioned, just that he
is able to observe in this way—so there is no "privileged" position of
observation, but it is necessary to be able to observe a certain amount
more than just the guy on the train or the guy on the tracks.

If you add that there are some absolute statements that can be derived
from these observations, you get a more complex statement of
"relativism"—whereby you understand that although things appear
differently from different perspectives, certain things do remain the
same, and the task of social science is to find out what those things are.

This is what cultural relativism in the old-fashioned anthropological sense meant—that there are certain codes of rationality and internal consistency that hold in all cultures, regardless of how odd or irrational their views might appear from the outside.

It's really a very formalist argument as I read it. Latour wants to suggest translating the formal properties of Einstein's argument into social science, in order to see both what social scientists can learn about "society" and how they use the term, and what hard scientists can learn about their own assumptions. He is trying to explain relativity only insofar as he wants to come up with a formal ("semiotic") reading of it that can be transferred to society. He's looking for a model for understanding social reality that will help social scientists deal with their debates—which have to do with the position and significance of the observer, with the relation between "content" of a social activity and "context" (to use his terms), and with the kinds of conclusions and rules that can be extracted through observation.

Since the questions in the field are often rather fuzzy, the argument is a bit vague and suggestive. However, I read it as a corrective aimed much more at sociologists than at scientists—not as an attempt to explain relativity to anyone, but as an attempt to pull out of the explanation of relativity offered in Einstein's essay certain useful ideas.

To this I add only that although Latour is not primarily interested in relativity as physics, there are passages in which he gets this aspect not only right, but eloquently so:

Instead of considering instruments (rulers and clocks) as ways of representing abstract notions like space and time, Einstein takes the instruments to be what *generates* space and time. Instead of space and time being represented through the mediation of the instruments, it is space and time which have always been representing the humble and hidden practice of superimposing notches, hands, and coordinates. It must be said that the character portrayed by Einstein does a very similar job to that of an anthropologist of science who refuses to understand what "space" and "time" mean, and who focuses instead on work, practices, and instruments. Like any constructivist in sociology of science, Einstein's first move in this text is to bring the abstractions back to the inscriptions and to the hard work of producing them.

Latour's first two sentences provide an exemplary encapsulation of the essential core of relativity. He then draws a parallel between Einstein's deconstruction of the notions of space and time, and the approach of social scientists to the content of science. It was indeed a convention among scientists, buried so deep in their culture as to be unrecognizable as such, that space and time were real objective entities measured by clocks and meter sticks. Einstein's profound insight was that, on the contrary, space and time are abstractions, serving to coordinate the results of such measurements. This is what sociologists of scientific knowledge have been saying for over two decades about all kinds of entities that scientists view as objectively real. We may not like it when the analogous thing is said about "the electron," but looking at such claims from this perspective transforms them from absurdities into serious suggestions, deserving not Olympian mockery, but reasoned debate.

Did I cheat by picking out the one coherent paragraph in the essay? I don't think so. Read it yourself. There are, to be sure, many obscure statements that appear to be about the physics of relativity, which may well be misconstruals of elementary technical points. But they are peripheral to the central issues, and some current diatribes are superficial in their identifications of "error." Latour, for example, appears to use "frame of reference" indifferently to whether he has in mind the position or state of motion of an observer. But this is irrelevant to his analogy between the Einstein character and anthropologists, and as Liz's comments reveal, the word "position" is used in cultural studies to mean conceptual location or ideological stance.

Latour has also been taken to task for maintaining that you need three frames of reference, rather than merely two, to make sense of the whole business. If you publicly condemn a fellow scholar for error, however, you ought to make sure you've got it right yourself. The fact, not widely appreciated, is that if you want to extract the Lorentz transformations without using Einstein's second postulate, then although you cannot do it using only two frames of reference, you can if you introduce a third [4]. Putting it formally, you must not only require that the inverse of a Lorentz transformation is a Lorentz transformation, for which the familiar two frames suffice, but also that they form a group: the product of two is a third. This requires a less familiar consideration of three frames to establish. While Latour clearly has something quite different in mind—two cultures and an anthropologist—if you're going to read him as getting the physics wrong, you should take care to get it right yourself.

I recommend two principles to guide what we scientists say and write in our exchanges with science critics. First, assume, at least as a preliminary working hypothesis, that you are reading intelligent people trying to make serious points, writing within a literary tradition that is as technical and unfamiliar to you as the professional idiom of your science may be to them. Second, technical criticisms should be based on reasoned argument. *Ex cathedra* sneering at selected sound bites demeans scholarly debate and is unlikely to persuade the sneered-at of one's serious intent. Try to think imaginatively about the rather subtle constellation of issues that may becloud superficially obvious "refutations" of "error."

Some of the attacks on science critics suggest a Germanic philologist scrutinizing Mark Twain's hilarious essay "The Awful German Language," where appear the immortal words, "he would rather decline two drinks than one German adjective."

"Ah," says the scholar, "here Mr. Twain, by the ludicrous error of using 'decline' in its colloquial sense of 'turn down,' reveals his abysmal ignorance of elementary grammatical theory." Some of the shots the science side has been firing in the science wars are hardly more accurate. The science critics get many things wrong, but we have to take more care explaining why, or we will only lend further credence to some of their worst misreadings of what we are up to.

References

1. B. Latour, *Social Studies of Science* 18, 3 (1988).
2. A. Sokal, *Social Text* **46/47**, 217 (1996).
3. S. Weinberg, *New York Review of Books*, 8 August 1996, p. 11.
4. See, for example, Y. P. Terletskii, *Paradoxes in the Theory of Relativity*, New York: Plenum (1968), sec. 7.

20. Postscript

1. *Physics Today* published three critical letters to the editor. Every one of them took up my quotation from Latour about instruments generating space and time. The first said that statements like Latour's "are thick upon the ground." It took the phrases I praised to be banalities. The second said that Latour's phrases were mistranslations from Einstein's German. It took the phrases to be wrong. The third said that "When Mermin praises Latour for asserting that in relativity 'Einstein takes the instruments to be what generate space and time,' the science wars have already been lost." It took my praise of Latour to be the end of civilization as we know it.

2. My giving a third of my column over to my "uniquely qualified" daughter was both misinterpreted and disapproved of. Since one of the complaints about essays like Latour's was that people in cultural studies would be misled into thinking they were being instructed in physical science, it seemed to me important to provide an example of how Latour was actually read by such a person. That a few people took literally my spoof on irrepressible parental pride reinforced my point that in matters like

these some physicists have taken leave of their sense of humor. I discovered on a blog a further reference to me and my daughter: the reason for my otherwise inexplicable view of Latour and his critics was to be found in a personal conflict of interest.

3. I received a letter from Bruno Latour. He said that he supposed that he ought to thank me for coming to his defense. But I had entirely failed to understand what he was getting at. So, he added, had my daughter.

4. Happily, the science wars are over. It took George Bush (the Second) to make it clear to all sides what was the real menace to science. As a clear indication that peace had arrived, in 2008 Bruno Latour was elected to that bastion of intellectual respectability, the American Academy of Arts and Sciences. As far as I know no Fellows of the Academy resigned in protest.

21

How not to create tigers

O ew'ge Nacht! Wann wirst du schwinden? Wann wird das Licht mein Auge finden?

–Tamino

Long-time readers of *Physics Today* may recall a series of conversations with my opinionated friend and colleague Professor Mozart, reported in *Reference Frame* columns early in this decade (Chapters 8, 9, 11, 12, and 13.) In answer to inquiries about his long silence, I can now reveal that Mozart mysteriously disappeared at the height of the Gingrich revolution in late 1995, demoralized by the growing obsession with "strategic research." I have just learned and am happy to report that he is alive and well, the proprietor of a small tobacco plantation in central Connecticut.

I had transcribed one of my last conversations with Bill Mozart in early 1995, but before I could negotiate his permission to report it in these pages, he vanished without a trace. As soon as I rediscovered his whereabouts, I sent him a brief note of inquiry, enclosing the text reproduced below, and was delighted to get it back decorated with the familiar scrawl I had despaired of ever seeing again: "PT-OK-WAM." I am publishing our four-year-old conversation today both for the insight it sheds on the state of mind that led so productive a physicist to drop out of the profession at so early an age, and also because Professor Mozart's views on the state of our discipline in the mid-1990s remain relevant to the difficult situation in which we find ourselves today, at the brink of the new millennium.

A little background: Several years before the conversation reported below, Professor Mozart had become deeply involved in satisfying the congressional demand for international assistance in the construction of the Superconducting Super Collider. In seeking contributions from abroad he had given free range to his prodigious imagination, and for several years after the cancellation of that visionary undertaking, he was fully occupied with the unwinding of his far-flung operations. His normally sunny disposition was clouded by his ongoing dismay at the failure of the great dream.

To aid younger readers in understanding what follows, I should also explain that well into the mid-1990s there were people—even a few in the State Department—who remembered that Russia, Ukraine, Belarus, Georgia, and many other countries once constituted a single nation that claimed to be "building" something called "communism." The disintegration of this effort was still on Mozart's mind, as my record of our last meeting reveals.

***

For the first time in years, I could sense cigar smoke as I entered the building, and by the time the elevator let me out on the fifth floor the scent of fine Havanas was unmistakable. His office door was ajar, so I went right in and there he was, clearly visible across the room through the haze, struggling to wrap an Oriental carpet in enormous sheets of brown paper. "W. A.," I cried out, "where on earth have you been?"

"Everywhere on Earth," he sighed. "I've returned cigars to Cuba, carpets to Kazakhstan, peanuts to Senegal. Sicilian carts for hauling magnets in the tunnel, crates and crates of Chinese seminar tea. They wanted international support? I got them international support! Not enough ... not enough ... ." he trailed off, emitting a fresh jet of smoke. "So," he resumed, perking up a bit, "I've taken it all back. Everything. Overlooked this carpet," he muttered, tying one more knot with a great flourish, "but now it's all accounted for. Every jot and tittle."

"The cigars?" I ventured, timidly. "Private side order," he shot back. "Everything is now returned. And we can start asking ourselves what next."

"In the import-export business?" I ventured.

"In the pursuit of an understanding of nature," he corrected me, more in sorrow than anger, "in the post-SSC era."

"You seem to have taken the cancellation rather hard," I observed, "for somebody who once called the study of elementary particles the archeology of physics."

"And so it is," he noted. "But archeology is a noble pursuit. The only comfort in the whole business is that the discoveries will still be there, waiting to be made, when we finally manage to summon up the will to make the attempt. No danger of earthquakes pulverizing everything or tidal waves washing it all away. So *sub specie aeternitatis* little has been lost if we have to wait for the 22nd or 23rd century to learn whether the Higgs is really there or merely a figment of an overactive 20th-century imagination. The main problem is going to be keeping the flame of learning alive over that long gap, so that when mankind finds itself again ready to resume the quest, there will still be a few who know what it's all about and can rekindle the fire in others."

"So you plan to retire to your study and write books?"

"Certainly books will have to be written, but books alone will not be enough. Were professional baseball to undergo a similar collapse, there would have to be a complete codification of the underlying principles for there to be a hope of reviving the game in a later century. But that would not suffice. In isolated villages and towns, we would need small bands of volunteers quietly working hard to maintain their skills with ball, bat, and glove, passing a working knowledge of their craft on to a younger generation who in turn would struggle to refine their capabilities and transmit them to yet the next generation, waiting, hoping, until the time came to resume the game at the highest level.

"In physics we don't even know the complete rules of the game, and probably never will. And the rules that we do know are so strange that were we merely to write them down and close the shop for a few generations, nobody would be able to make any sense of them when the time came to open the book. If we don't stay in training, in a generation or two nobody will be able to make any sense of it. Which means that the universities are now more important than they have ever been before."

"You mean ...?"

"Yes, to keep learning alive. As the monasteries preserved literacy during the Dark Ages, so must the universities bend to the task of keeping learning alive during the dark night of strategic research that now sweeps irresistibly over us." He was wracked by so violent a sob that I felt obliged to try to put a cheerful spin on the vision he seemed to find so grim,

"So that when the fruits of strategic research have brought prosperity and abundance to all, we can resume the real quest?" I cheerily injected.

"No," he sighed, "so that when the sterility of attempting surgically to extract golden eggs from flocks of anesthetized geese becomes evident even to the most avaricious, we can still have a hope of averting the impending disaster..."

"By resuming the search for the Higgs?"

"By restoring the climate in which free-range geese can lay their golden eggs unmolested in the tall grass."

I tried to dislodge him from his avian metaphor. "You are waiting for the day when the nation is again prosperous enough to indulge in the luxury of curiosity-driven research?"

A shudder of revulsion swept over him. "A phrase that calls to mind little boys and girls playing doctor—probably as close to research as the thinker who coined it ever got. No, I am awaiting the day when people remember the fact that discovery does not work by deciding what you want and then discovering it. Until that happens, we are trapped in an epoch of stifling creationism."

"What do you mean, creationism? Nobody is suggesting that the physical world requires no explanation beyond a declaration that God made it that way. At least nobody in Congress is. At least not yet. Are they?" A wave of alarm swept over me. With this remark, I seemed to have finally reached his great reserves of pedagogical energy. He smiled at my foolishness. "The problem with creationism is not just that it appeals to God to obviate the need for thought. If that were all, it would be easily disposed of. But today, though creationism has suffered a crushing defeat in the former Soviet Union, it is now running rampant in the United States."

"You've lost me, W. A. How was the collapse of atheistic communism a crushing defeat for creationism?"

His good humor thoroughly restored, Professor Mozart gave me a tolerant smile. "Religion offers just one avenue for creationism to do its mischief. Communism was creationism in the economic sphere." Noting my blank expression, he elaborated. "Creationists assume that the passage from the actual or conjectured starting point to the actual or desired final stage has been or can be achieved by direct and purposeful construction of the latter from the former."

The light dawned on me. "You mean religious creationists maintain that we got from the void to an Earth teeming with life through the purposeful intervention of God?"

He smiled and moved as if to pat me on the head, ending up offering me a cigar instead.

"And," I went on breathlessly, declining the offer with a shake of my head, "economic creationists maintain that we go from an inequitable distribution of wealth to a just society by the purposeful action of government?"

His smile seemed to light up the room.

"And the proponents of strategic research are creationists because …" I ran out of steam and sheepishly grinned back at him.

Professor Mozart rose to his feet. Waving his cigar like a conductor with a baton, he declared, "Proponents of strategic research hold the creationist belief that science works by deciding where we want to get and then going there. They would prescribe for science what Stalin prescribed for agriculture. It will have a similar success."

"What's wrong with going where you want to get to?" I foolishly inquired.

"What's wrong with it?" Mozart thundered, "what's wrong with it? Nothing," he hissed, suddenly pianissimo, "if where you want to get is to the mountains for a hike or the delicatessen for a sandwich. But it doesn't make very much sense to insist on where you want to end up when you're dealing with a process that always manages to get you somewhere wonderful, provided only you don't specify in advance the particular nature of the next miracle."

"How can you get anywhere interesting if you renounce any advanced planning?" I asked.

"Are tigers interesting?" he snarled at me.

I jumped back a good two feet. "Yes," I said, having always liked zoos.

"So you maintain that the first thing that crawled out of the primeval slime sat down and drafted a proposal to the National Institutes of Health for a grant to help it become a tiger?"

"Of course not," I replied. "Natural selection led in the course of time to an amazing variety of more and more complex forms, the particular character of the complexity of each variation being governed entirely by chance. Planning had nothing to do with it."

"Right," he pronounced. "No immortal hand or eye could frame a strategic plan for that fearful symmetry. The only requirement was for the surroundings to be both challenging and hospitable enough for wonderful things to develop. One of them happened to be a tiger. Another, an oak tree. Yet another, you. All wonderful, to be sure, but all unplanned and all intrinsically unplannable."

And fixing his eyes on me as if I were a specimen of a possibly interesting species of butterfly, he completed the last knot in the last loop of cord, hoisted the fully wrapped carpet over his shoulder, and headed for the door. "They'll be glad to get this beauty back in Kazakhstan," he sang out.

"What will you do when you've shipped it off?" I called after him.

"Think," the answer drifted down the hall. "Sit back and think."

"What about?" I shouted.

"If I knew that," he proclaimed, as he disappeared into the freight elevator, "there'd be little point in doing it!" And as he slowly descended to the loading dock his parting benediction for science drifted up the shaft:

Tyger! Tyger! burning bright

In the forests of the night ...

21. Postscript

1. This grew out of an old Professor Mozart draft, which had sat incomplete in my files for almost five years. When I rediscovered it I was particularly struck by Bill Mozart's view that creationism was not just a foolish response of religious fundamentalists to the threat they perceived from evolution. It was a menace to thinking and planning that had plagued humanity throughout the 20th century.

2. I also wanted to record Mozart's continuing devotion to the aborted Superconducting Super Collider.

3. And I was moved by some Mozart fans who lamented his long absence from the pages of *Physics Today*.

22

What's wrong with this elegance

A little while ago I was asked to give a lecture at the very elegant Weisman Art Museum in Minneapolis on an assigned title: "Elegance in Physics." As I get older, the things I'm asked to do get stranger, so I wasn't surprised. Alarmingly, the older I get, the stronger is my inclination to do the peculiar ones. So I accepted the invitation, and soon found myself brooding, not, as I had imagined, about the glory of the eternal verities, but about the highly contentious nature of elegance in physics.

Here is the first such difference of opinion I came upon. In a lecture at Fermilab with a title similar to the one I was given, Subrahmanyan Chandrasekhar talked about "harmoniously organizing a domain of science with order, pattern, and coherence."[1] He cited five examples of such pinnacles of exposition, one of them being Paul Dirac's celebrated book, *Principles of Quantum Mechanics*. "The translucence of the eternal splendor through material phenomena," Chandrasekhar remarked, "[is] made iridescent in these books."

Keeping that iridescent translucence firmly in mind, consider the following remarks of the eminent mathematician Jean Dieudonné:

When one arrives at the mathematical theories on which quantum mechanics is based, one realizes that the attitude of certain physicists in the handling of these theories truly borders on delirium ... One has to

---

1 See his article in *Physics Today*, July 1979, p. 25.

ask oneself what remains in the mind of a student who has absorbed this unbelievable accumulation of nonsense, real hogwash! It would appear that today's physicists are only at ease in the vague, the obscure, and the contradictory [1].

What is Dieudonné talking about? He is addressing the approach to quantum mechanics laid out in Dirac's book.

Elegance in physics is as much in the eye of the beholder as it is in any other field of human endeavor. Dirac's formulation appeals to physicists because, by being a little vague and ambiguous about its precise mathematical structure, it enables them to grasp and manipulate the physical content of the theory with a clarity and power that would be greatly diminished if one were distracted by certain complicating but fundamentally uninteresting mathematical technicalities. But for mathematicians, those minor technical matters lie at the heart of the subject. Quantum mechanics becomes ill-formulated and grotesque if it does not properly rest on impeccable mathematical foundations.

Chandrasekhar and Dieudonné having thus sensitized me to what one might call interdisciplinary aesthetic dissonance, I realized that the same mechanism had been at work in a difference of opinion I had had a few years ago with crystallographers.

For over a century, the hallmark of crystallinity had been taken to be periodicity at the atomic scale. All crystals were thought to be built out of primitive units made from a comparatively small number of atoms, repeated over and over again. But then new crystalline materials were discovered. Some of them, dubbed quasicrystals, had symmetries that no periodic structure could possibly have. One of the important features all these aperiodic crystals shared with the ordinary periodic ones was a characteristic display of sharp Bragg peaks in their X-ray photographs.

The crystallographers realized early in the game that all these aperiodic crystals could be described as three-dimensional cross sections of structures periodic in spaces of four, five, six, or even more dimensions. Motivated by this insight, the International Union of Crystallography has had for many years a subcommittee that struggles over the appropriate nomenclature to describe the kinds of symmetries one can encounter in four or more dimensions. Thinking about four-dimensional crystals can be highly entertaining, and is it not elegant that nature has actually produced materials that require us to think that way?

No, it is not. The new materials don't require us to think that way. Some of us noticed that if we shifted the defining feature of crystallinity from the periodicity of the atomic structure to the presence of Bragg peaks in the diffraction diagram, then the only relevant symmetries continued to be those associated with patterns of reflected X-rays in boring old three-dimensional space. The patterns made by aperiodic crystals can be more complex, but the geometrical description of the symmetry of those patterns remained familiarly three-dimensional.

We thought that this triumphant return to three-dimensional geometry, at the price of shifting the emphasis from the crystal itself to its X-ray photographs, was an elegant step forward. We expected the crystallographers to throw out their irrelevant books on higher-dimensional geometry, dissolve or at least redirect their commissions on higher-dimensional nomenclature, and glory in the elegance of rethinking crystal symmetry in terms of diffraction patterns.

Did that happen? No, of course not! Most crystallographers found our beautiful and illuminating shift from the structure of the crystal to the structure of its diffraction pattern to be unnatural, ungainly, and unintelligible. They were not impressed that our approach provided a direct link between crystal symmetry and electronic properties, because, being crystallographers but not physicists, they were not interested in electronic properties. I won't abuse my privilege as a *Reference Frame* columnist to elaborate on why our way is better, but I report the sad tale here as another, less lofty example of the relativity of elegance.

With two such examples to think about, it occurred to me that a curious episode, early in my professional career, was yet another manifestation of the same phenomenon.

Over 25 years ago, I became interested in the physics of the newly discovered superfluid phases of helium-3. I realized that one of the phases of this unique fluid bore a striking similarity to a type of structure known to mechanical engineers as a Cosserat continuum. So when I noticed one day that there was a seminar on Cosserat continua on the Cornell Engineering quad, I wandered over.

I didn't learn anything useful about helium-3 from the engineers, but in the discussion period after the lecture, rather to my surprise, a heated debate broke out over whether a point particle could have an angular momentum—the terms of the argument were whether a particle with no internal structure at all could nevertheless spin like a top. I found this dispute remarkable for two reasons. First, because I hadn't thought that hard-headed engineers could become so passionate about so fundamentally metaphysical an issue. And second, because the question, insofar as it had empirical content, had an elegant answer whose relevance to their argument the disputants seemed not to have noticed. So I rose to my feet and made a remark, elegantly stated in four words, that I was sure would settle the whole debate: "What about an electron?"

There followed a sickening silence. It was as if someone in the crowd had shouted an obscenity. (This was the early 1970s, when somebody in any crowd was quite likely to shout an obscenity.) A senior professor of theoretical and applied mechanics rose slowly from his seat, fixed me with a baleful gaze, and delivered this crushing rejoinder: "Have you ever *seen* an electron?" His riposte elicited nods and murmurs of approval throughout the auditorium. Then they returned to their deliberations with undiminished vigor.

My elegant invocation of physical reality to cut through a metaphysical argument was viewed as a clumsy introduction of speculative metaphysics into a tough-mindedly practical debate about—about what? To this day, I do not know what the debate was about. So I slunk back to the physics corner of the campus, where the elegance and relevance of spinning electrons remained unchallenged.

Having thus become well attuned to the highly contentious nature of elegance in physics, I now realize that an excellent example of debatable elegance is provided by the new field of quantum computation, which I offer as a final illustration of the contingency of scientific aesthetics.

You can set up a quantum computer to act on a superposition of all possible inputs. Because a system in a superposition of inputs evolves into the superposition of the outputs it would have evolved into for each of the superposed possibilities, in no more time than it takes to do a single calculation, a quantum computer with an n-bit input register can produce a state whose structure encodes the outcome of 2^n separate calculations! If you have 100 bits—hardly anything for a classical computer—that amounts to doing $2^{100} \approx 10^{30}$ calculations in a single pass.

But have you really *done* that astronomical number of calculations? How much of that vast output of information can actually be extracted? Not much! Indeed, the most obvious approach gives you only a single output, randomly chosen from the enormous range of possibilities, so you do no better than you would have done by feeding a randomly chosen input into a classical computer. But one of the funny things about quantum mechanics is that it offers you trade-offs. If you're willing to renounce the possibility of getting any information about any individual computation, you can get certain kinds of partial information about the results of all the computations. In particular, if you know that the output is a periodic function of the input but you don't know the value of the period, then a quantum computer turns out to offer you clues that permit you to determine the value of the period spectacularly more efficiently than you can with a classical computer.

Is this elegant? It is for computer scientists, because it offers a striking demonstration that computational complexity theory—the study of how the time it takes to do a computation scales with the size of the input—cannot be divorced from assumptions about the physical nature of your computer. It is elegant for the National Security Agency, because finding the period of an unknown periodic function is the key step in cracking a widely used coding scheme. For me the elegance lies in the entirely new perspective quantum computation provides on the exquisitely intricate ways in which information can be encoded in quantum states.

On the other hand, many people find it altogether inelegant, because of the enormous, quite possibly insuperable (but please don't tell the NSA!) obstacles in the way of building a quantum computer capable of performing these wonderful tricks.

Quantum computation raises the question of whether feasibility is or is not an essential ingredient in determining the elegance of a proposed technology. Ordinarily it surely is, but quantum computation seems to me a case where the conceptual charm of the idea is so very powerful that the strong possibility that it will never prove feasible fails to undermine its elegance. Long may it flourish, if only as a *gedanken* technology!

Reference

1. J. Dieudonné, in *La Pensée physique contemporaine*, S. Diner, D. Fargue, G. Lochak (eds.), Paris: Editions Augustin Fresnel (1982).

22. Postscript

1. A letter to *Physics Today* solved a mystery that had puzzled me for years. Boltzmann's famous brusque dismissal of the role of elegance in science, "Elegance is for tailors," is widely attributed to Einstein. Why? I learned from the letter that Boltzmann's

pithy rejection of elegance got attributed to Einstein because Einstein cites the quotation in the preface to the very same little popular book on relativity that inspired Bruno Latour to produce the essay that I discuss in Chapter 20.[2] Einstein explicitly attributes the statement to Boltzmann, but that was no match for the power of the Matthew effect.[3]

2. Another letter came from one of the crystallographers whose views on aperiodic crystals continued to be favored in spite of the more elegant formulation put forth by me and my colleagues. He maintained that our failure to convince people that our way was better demonstrated that his way was indeed more elegant. Fair enough!

2 I say more about misattributions in Chapters 25 and 26. I have more to say about the Einstein–Boltzmann mix-up in Chapter 37.

3 For more on the Matthew effect see Chapter 26.

23

The contemplation of quantum computation

At the heart of the puzzlement induced in many by the quantum theory lies a tension between reality and knowledge, between facts and information. Do the basic entities of the theory, quantum states or their wavefunctions, directly correspond to something in the real world so that something—wavefunction stuff—actually does pass through both slits in a two-slit interference experiment? If so, then either we are faced with the puzzle of how something real can suffer abrupt changes in response to faraway events or, if we banish wavefunction collapse from the theory, we are faced with a reality that absurdly evolves to encompass myriads of alternative histories growing ever more unalike. These puzzles melt away if the basic entities of the theory are merely representations of knowledge, only to be replaced by others. Whose knowledge? Knowledge of what?

The new *gedanken* technology of quantum computation provides an unfamiliar perspective on such vexing questions, by using the quantum theory, not to expand our understanding and control of the physical world, but to exploit the quantum behavior of the physical world as a novel way to encode and process information. The information is primary; the underlying physical system only matters as a vehicle for that information. Quantum computer scientists view a set of n interacting spins-½ not for the insight it offers into the nature of magnetic materials, but as a way to represent and manipulate integers, through their n-bit binary representations as orthogonal states in a "computational basis" that specifies whether each individual spin is up (1) or down (0).

Quantum computation differs in several crucial ways from ordinary classical computation:

▶ The states of a quantum bit (which I here call a Qbit, in quixotic defiance of the fashionable but orthographically preposterous "qubit") are not restricted to 0 or 1, as in a classical computer, but can be in arbitrary superpositions of 0 and 1.

▶ Even more foreign to classical intuition, the state of the entire computer can be what Schrödinger called *verschränkt* (entangled), with individual Qbits having no (pure) states of their own at all. While a classical computation flips classical Cbits between 0 and 1, a quantum computation subjects the Qbits to the much more general unitary transformations that specify the time evolution of quantum states.

▶ At the end of that unitary evolution, one cannot read out the final state of the Qbits, as one can with classical Cbits; the only way to extract information is to make a measurement.

In designing a quantum computation taking the N integers from 0 to $2^n - 1$ into themselves, one invariably works with $2n$ Qbits. The first n Qbits (the "input register") can initially represent a number x between 0 and $N - 1$ and the second n (the "output register") are initially all zero. After an appropriately designed unitary time evolution (but before the final measurement), the input register still contains the same number x, but the output register contains the result, $f(x)$, of the calculation.

A neat thing about this superficially pedestrian arrangement is that one can easily put the input register into an equally weighted superposition of states representing *all* the integers between 0 and $N - 1$. (Starting with the state that represents $x = 0$, with every spin down along the z direction, and simply rotate each spin by 90° about the y-direction.) Because the unitary time evolution is linear, the output is guaranteed to be an equally weighted superposition of product states in which each of the $N - 1$ possible states $|x\rangle$ of the input register is paired with the state $|f(x)\rangle$ of the output register:

$$|\Psi\rangle = \frac{1}{\sqrt{N}} \sum_{x=0}^{N-1} |x\rangle |f(x)\rangle. \qquad (1)$$

If the number of Qbits in our quantum computer is a mere 200—not enough Cbits to store this clause in a classical computer—then specifying the state $|\Psi\rangle$ requires evaluating $f(x)$ for $2^{100} \approx 10^{30}$ different x. Yet except for the trivial extra step of producing the initial state, the pre-measurement stage of the computation takes no longer than a single evaluation. This leads some practitioners of quantum computation to wax poetic. "Where," they inquire, "were all those evaluations calculated?" "In a million trillion trillion parallel universes!" they joyfully answer. And indeed, if you believe the quantum state has an objective existence like a classical field, it is hard not to fly off with them to cloud-cuckoo land.

But of course one can turn the poetic argument upside down, regarding the mere possibility of a machine ending up in a state like $|\Psi\rangle$ as powerful evidence that a quantum state does not exist as anything objective, but merely encodes what information can be extracted from a system described by such a state. What is not a matter of interpretive debate is that given a computer in the state $|\Psi\rangle$, a measurement can reveal only extremely limited information about what that state actually is, and therefore only highly restricted features of the results of those gazillions of evaluations.

Indeed, the most obvious way to get information from $|\Psi\rangle$, by measuring both input and output registers in the computational basis, simply gives $f(x_0)$ for a random choice of x_0, revealing no more than could have been learned in a single evaluation of f. To be sure, that random choice of x_0 is made *after* the computer has done the bulk of its labor, rather than before, which sounds deliciously strange. But since we have absolutely no control over which x_0 is selected, this works no differently from making the random choice before the calculation.

There are, however, more subtle ways to extract information from a quantum computer. Suppose, for example, we have a function f

with only two inputs ($x = 0$ or 1) and two outputs ($f(x) = 0$ or 1). Behind this simplicity could lie an enormous computation, carried out by a quantum subroutine. For example, $f(x)$ could be zero or one depending on whether the millionth bit of the binary expansion of $\sqrt{2+x}$ was even or odd. Using this quantum subroutine, we can build a quantum circuit that takes the initial state $|x\rangle|0\rangle$ of the single Qbit input and output registers into $|x\rangle|f(x)\rangle$. Since unitary time evolution takes orthogonal states into orthogonal states, the quantum circuit must also take $|x\rangle|1\rangle$ into $|x\rangle|\overline{f}(x)\rangle$ (except for an unimportant phase), where $\overline{f} = 1 - f$.

The obvious thing to do with such a device is to learn the value of either $f(0)$ or $f(1)$, by letting it act on $|0\rangle|0\rangle$ or $|1\rangle|0\rangle$ and then measuring the output register. As noted above, if we are in a gambling mood, we can instead prepare the input register in the superposition $(|0\rangle+|1\rangle)/\sqrt{2}$, run the computation, and then measure both the input and the output registers to have a 50–50 chance of learning either $f(0)$ or $f(1)$ (as well as learning which of the two we have learned).

Suppose, though, that we want to know only whether or not $f(0) = f(1)$: Are the millionth bits in the binary expansions of $\sqrt{2}$ and $\sqrt{3}$ the same or different? With a classical computer it's hard to imagine doing anything but calculating $f(0)$ and $f(1)$ and comparing the answers, requiring two invocations of the subroutine. But with a quantum computer we can do better. We can put *both* input *and* output registers into the superposition $(|0\rangle-|1\rangle)/\sqrt{2}$. The joint state of the two registers is then proportional to

$$|\Psi\rangle = |0\rangle|0\rangle - |1\rangle|0\rangle - |0\rangle|1\rangle + |1\rangle|1\rangle \qquad (2)$$

and because time evolution is linear, the computation simply takes $|\Psi\rangle$ into

$$|0\rangle|f(0)\rangle - |1\rangle|f(1)\rangle - |0\rangle|\overline{f}(0)\rangle + |1\rangle|\overline{f}(1)\rangle. \qquad (3)$$

Depending on whether $f(0)$ and $f(1)$ are or are not the same the final state (3) of the two registers can be written either as

$$\left(|0\rangle-|1\rangle\right)\left(\left|f(0)\rangle-|\bar{f}(0)\rangle\right)\right), \quad f(0)=f(1), \qquad (4)$$

or as

$$\left(|0\rangle+|1\rangle\right)\left(\left|f(0)\rangle-|\bar{f}(0)\rangle\right)\right), \quad f(0)\neq f(1). \qquad (5)$$

Notice that now the input register is in one of two orthogonal states—eigenstates of the spin component σ_x—depending on whether $f(0)$ and $f(1)$ are or are not the same. By measuring σ_x on the *input* register we can learn whether the millionth bits of $\sqrt{2}$ and $\sqrt{3}$ are or are not the same with only a single invocation of the quantum subroutine!

But we have had to pay a price. Regardless of what we learn from measuring the input register, and regardless of the actual value of $f(0)$, the output register will be left in exactly the same state. So we can learn precisely nothing about the actual value of $f(0)$ or $f(1)$. By cleverly setting up our computer so a single call of the subroutine tells us whether the millionth bits of $\sqrt{2}$ and $\sqrt{3}$ are the same or different, we have lost the ability to learn from that single use of the subroutine what those bits actually are.

This smells like the archetypal quantum trade-off. You can learn the position of a particle if you renounce the possibility of learning its momentum, and vice versa. We learned many such examples in childhood. But we also learned that the answer to the question we chose not to ask was not only lost forever, but never existed in the first place. The information we chose to acquire was not about some pre-existing condition ("knowledge of what?"), but an inseparable component of the very act of inquiry.

Here the quantum computation is interestingly different. When the physical process represents a computation, there is nothing problematic about the character of the complementary kinds of knowledge we have to choose between: It is all knowledge about arithmetic. Even the most anti-Platonist of mathematicians would acknowledge that all the answers *exist*, whether or not anybody has ever taken the trouble to find out what they are. The situation is quintessentially quantum except that everything—both what we learn and what we must renounce learning—*exists* whether or not we have learned it. The millionth bit of $\sqrt{2}$ *does* have a value, even if nobody has ever looked at it.

While a cleancut resolution of the tension between the real and the known has yet to emerge from the flourishing explorations of quantum computational algorithms during the past half decade, new *gedanken* applications of quantum mechanics like the one I have just described offer fertile ground for reconsidering the knowledge-vs-reality muddle we have been thrashing about in for the past 75 years. It would be interesting, but maybe a bit embarrassing, if the new breed of computer scientists were the ones to straighten us out. So I commend quantum computation to those young philosophers and old physicists who have the time and inclination to think about these issues. It's by far the most interesting such game currently in town.

23. Postscript

1. This essay on quantum computation is more technical than most chapters, but the issues it addresses are primarily philosophical.

2. The strikingly different way complementarity plays out in a quantum computation, described in the three final paragraphs, is accounted for by the QBist view of quantum mechanics (Chapters 31–33). Complementarity is not about what might or might not exist. It is about the personal experiences that a user of quantum mechanics might or might not have.

24

What's wrong with these questions

On 15 August 2000, *The New York Times*, celebrating the new century, published a list of 10 questions that they characterized as ones physicists would like to ask their colleagues in the year 2100 if they awoke from a hundred-year sleep.

1. Are there reasons why the fundamental dimensionless parameters have the values they do?
2. What role did quantum gravity play in the Big Bang?
3. What is the lifetime of the proton?
4. Is supersymmetry a broken symmetry of nature?
5. Why is spacetime apparently four-dimensional?
6. What is the value of the cosmological constant, and is it really constant?
7. Does M-theory describe nature?
8. What happens to information that falls into a black hole?
9. Why is gravity so weak?
10. Can we quantitatively understand quark and gluon confinement?

You will not be surprised to learn that the questions were assembled at a party celebrating the conclusion of a conference on superstring theory. The *Times*, however, characterized them as "Physics questions to ponder," leaving me to ponder why they were so different in character from what I would be most eager to learn from my professional descendants at the end of a hundred-year nap.

The *Times* inspired me to put together my own list of the questions I'd put to a colleague in 2100. The criteria for inclusion on my list are (a) that I would love to know the answer, (b) that the questions should be likely to make sense to scientists in 2100 and not just to historians of science, and (c) that the questions should have a reasonable chance of not eliciting titters at my early 21st-century naivety. Probably you'll find my list just as parochial as I found the string theorists' list. But here it is:

1. *What are the names of the major branches of science? What are the names of the major branches of physics, if physics is still an identifiable branch? Please characterize their scope in simple early 21st-century terms, if you can, or try to give me some sense of why my ignorance makes this impossible.*

   I can't imagine that the landscape will look familiar in 100 years. Already, for example—and this is bad news for physics—chemistry seems to be trying to become a branch of biology, as witnessed by the recent name changes of the departments at Harvard and Cornell, from Chemistry to Chemistry and Chemical Biology. Physics, on the other hand, seems bent on absorbing biophysics, defined in the broadest possible sense. Looking backward rather than ahead, what would a physicist from 1900 have made of the term "computer science," not to mention "information science"? You might complain that the real content of this question is "tell me everything of interest," but all I'd like to find out is what unfamiliar names are going to be there, and what familiar names are going to be missing.

2. *Please show me a widely used, widely affordable device that will astonish me in as many different ways as a laptop computer would have astonished a physicist in 1900. At least some of the purposes served by this device should be as comprehensible to me as the uses of a laptop computer would have been to a late 19th-century physicist.*

   When you think about it, the number of ways in which a laptop would have amazed in 1900 is itself amazing. Forget about its primary functions. What about the material its case is made of, its cost as a fraction of mean annual income, the source of its power, its ability to imitate a symphony orchestra, its ability to show glowing pictures that move? Nobody imagined it in 1900. Nobody today can imagine what extraordinary objects will be found in the households (assuming there still are households) or pockets (assuming there still are pockets) of 2100.

3. *Are fundamental theories still based on superpositions of states that evolve unitarily, or have the basic principles of quantum mechanics been replaced? If quantum mechanics has survived, have people reached a consensus on the solution to the interpretive problems, or have they simply ceased to view them as problems needing a solution? If quantum mechanics has not survived, has the theory that replaced it clarified these puzzles, or do people find it equally or even more mysterious?*

I worry that this question might elicit polite bewilderment. But an appropriate time scale for the survival of quantum mechanics is set by the fact that its basic conceptual machinery has suffered no alterations for three-quarters of a century. Even quantum field theory has as its most important application the calculation of cross sections, which exploits the same old unitarity of time evolution—the superposition principle—and the same old Born rule for extracting probabilities from states. So the persistence of the same formalism for another hundred years seems at least plausible.

If the theory is indeed still with us in essentially the same form in 2100, will there still be serious people, as there are today, who feel that in some fundamental sense we don't know what we're talking about? Or will the early 21st-century people who believed there ought to be a better way to understand the theory (if not the world itself) be assigned to the same dustbin of history as the early 20th-century ether theorists?

4. *Have intelligent signals of extraterrestrial origin been detected?*
   I hope somebody keeps on looking. And I hope, if the answer is still no, that our descendants don't regard the search as silly.

5. *Do time and space still play the fundamental roles they did in early 21st-century physics, or have they been replaced by more coherent, less ambiguous concepts?*

   How can people talk about spacetime turning into a foam at the Planck scale when we barely manage to define space and time at the atomic scale? Time, for example, is nothing more than an extremely convenient and compact way to characterize the correlations between objects we can use as clocks, and clocks tend to be macroscopic. To be sure, we can generate frequencies from atoms and correlate them with macroscopic clocks, but the shorter the length scale, the more it looks like you're talking about energies divided by Planck's constant. The connections with clocks become increasingly indirect. There seems to me to be a considerable danger here of imposing on an utterly alien realm a useful bookkeeping device we've merely invented for our own macroscopic convenience. Time and space will still be with us in 2100, but I'm not so sure they'll be in evidence at the foundations of the scientific description of nature, whatever that discipline happens to be called.

6. *Tell me about a collective state of matter, unimagined in the year 2000, that is as remarkable as, for example, superconductivity, superfluidity, or the fractionally quantized Hall effect seemed to be at the end of the 20th century.*

   Who could have imagined such phenomena in 1900? Surely the extraordinary capacity of bulk matter to behave in ways that transcend anything one could possibly have guessed from studying its constituents, will produce many comparably unimaginable things in the next 100 years.

7. *Are room-temperature superconductors an important part of your technology?*

This question might appear temporally provincial from the perspective of 2100, the recent flurry of interest in high-temperature superconductors being only 15 years old. But I am reassured by the fact that the broader quest has been with us now for almost a century, so it might not be presumptuous to guess that it could still be relevant in another hundred years. I'll take my chances that the question will not elicit giggles.One could ask a similar question for similar reasons about controlled nuclear fusion, even though that quest has only been with us for about fifty years.

8. *Has any progress been made in understanding the nature of conscious experience or how the mind affects the body, and does quantum mechanics or its successor play a fundamental role in that understanding?*

There are those who say consciousness is a nonproblem because the question doesn't make any sense, and those who say it is a nonproblem because the answer is obvious. Physicists further divide into those who say quantum mechanics clearly does or clearly does not have anything to do with it. The problem of consciousness, of course, has been around for many centuries. But the growing sense, at least among physicists, that science has something to say about it does not seem to me transparently absurd, even though no two scientists can currently agree on what that something might be. The titter risk here is substantial, but I'll take my chances. I'd love to know whether the question will be viewed as vexing, as silly, or as substantially answered by 2100.

9. *Did quarks turn out to be elementary or composite? If composite, did the candidates for their constituents turn out to be elementary or composite? Or do you have a better way of looking at these phenomena? What, indeed, is the lifetime of the nucleus of the hydrogen atom?*

If string theory is already a better way of looking at these phenomena, the question may be partly answered. Maybe a better way to phrase it is this: What energy scales have you been able to reach, and have you observed new structure all along the way, or have things finally started to simplify, as people once innocently expected that they would?

10. *Has anybody built a quantum computer that can factor a thousand-bit integer? What else is it used for? Do most homes have one?*

This is the most rash of my questions, since the whole subject of quantum computation is so new that it all may well have evaporated by 2010. The question would then make sense in 2100 only to specialists in the history of science. But I like to think that so beautiful a *gedanken* technology will capture enough imaginations to give the quest for its realization sufficient impetus that—who knows—it might even succeed despite all current indications to the contrary. I'm hoping my intellectual descendants in 2100 will at least know what it is I'm talking about. If so, this will probably mean they've succeeded.But I wouldn't bet my great-great-great-grandchildren's college tuition money on it. Assuming there still is money. Assuming colleges still charge tuition. Assuming there still are colleges.

24. Postscript

1. One eminent string theorist took my essay to be nothing more than an unjustified attack on the string theorists at whose conference the questions in the *New York Times* had been collected. To be sure, I had said[1] that the establishment of the e-print archive (now arXiv.org) "could well end up as their greatest contribution to science." But my real point was the extraordinary difficulty of anticipating the state of any human undertaking four generations in the future.

1 Chapter 12.

2. One speculation in *Questions* has already been settled. We are well past 2010, and interest in quantum computation has not faded away. But no serious quantum computer has been built.

3. Chapter 38 is an expanded version of *Questions*, the text of a lecture in Zurich in 2005, at a celebration of the centenary of Einstein's 1905 *annus mirabilis*. There I place the fuller discussion in a setting provided by Max Beerbohm's masterpiece, *Enoch Soames*.

25

What's wrong with this quantum world

If you ask Google to search for "no quantum world," you will get nearly 300 hits. They all give the following quotation (or recognizable corruptions of it):

There is no quantum world. There is only an abstract quantum physical description. It is wrong to think that the task of physics is to find out how nature is. Physics concerns what we can say about nature.

Over 90% of them attribute the statement to Niels Bohr, with phrases like "Bohr's dictum ...," "Bohr insisted that ...," "Bohr proclaimed ...," "Niels Bohr said, in a frequently quoted passage ...," "Niels Bohr *wrote* [my emphasis] ...," and even "Explain and evaluate Bohr's philosophy of quantum theory with reference to his assertion ... ."

Here is yet another example of the power of the internet to enrich our knowledge. There is only one problem. Bohr, who took writing very seriously indeed, never published such an assertion in any of his writings, although he repeatedly refined, reformulated, and often simply repeated his position on the philosophical foundations of the quantum theory.

The statement actually comes from an essay by Aage Petersen, "The Philosophy of Niels Bohr" [1], which he published in the *Bulletin of the Atomic Scientists* shortly after Bohr's death. Petersen introduced the words with

When asked whether the algorithm of quantum mechanics could be considered as somehow mirroring an underlying quantum world, Bohr would answer ...

So what may now be the most celebrated of all Bohr quotations on the nature of the quantum theory is at best an attempt by a close associate to characterize Bohr's general response to the highly problematic notion of a "quantum world," written too late for Bohr to respond.

When you ask Google to list only those pages that also mention the actual author of the words, Petersen, the number of hits drops from 286 to 18. The status of this "quotation" as hearsay is in danger of being lost.

I'm particularly sensitive to this risk because in the 1980s I used to enjoy giving physics colloquia on Bell's theorem and its implications for our understanding of quantum mechanics. I was always fond of Petersen's recollection of what Bohr used to say, and I felt that his formulation captured something important about the situation revealed by Bell's theorem. So I would read it aloud in those talks, always emphasizing that it was Petersen reminiscing, and not from anything Bohr himself had actually written.

That worked quite well until the day in 1982 when I gave the physics colloquium at MIT. To my great pleasure, Victor Weisskopf was sitting in his usual place in the front row, smiling approvingly up at me. (It's surprising how much such encouragement from such a source can improve the quality of a talk.) His smiles continued right up to the moment when I read the Petersen quotation. No sooner had I finished reading it than Viki was on his feet. The smile was now a frown. "That's outrageous," he proclaimed, "Bohr couldn't possibly have said anything like that!" Somewhat taken aback by this sudden flip from approbation to condemnation, I feebly protested that I wasn't attributing it to Bohr, merely to Aage Petersen's memory of Bohr. That did not extinguish the flames. "Shame on Aage Petersen," declared Viki, "for putting those ridiculous words in Bohr's mouth!"

(I must emphasize that although I have used quotation marks, as the rules of punctuation require, my Weisskopf "quotations" are based only on my unreliable memory of what Viki actually said 20 years ago. They are crude reconstructions that I hope capture the spirit of his remarks. If you like to collect Weisskopf quotations, please do not add these to your list.)

I have been a big fan of Weisskopf ever since, as a graduate student at Harvard in the late 1950s, I would watch him arrive at Harvard physics colloquia with the MIT contingent; he always made a point during the lecture of asking "dumb" questions. The purpose of the questions, as far as I could tell, was to help the students in the audience understand what was going on. His junior colleagues at MIT would invariably rush to provide patronizing answers, but that never bothered him or deterred him from the practice.

Since Weisskopf had spent considerable time with Bohr in the early days of quantum mechanics, I took his strong reaction quite seriously and dropped the Petersen quote from subsequent versions of my lecture. I did, however, comb the writings of Bohr to see if I could find anything that seemed to express a similar sentiment.

Was Bohr ever willing to publish anything like what Petersen said he used to say? The closest I can find to the Petersen pseudo-quotation is this:

Indeed from our present standpoint, physics is to be regarded not so much as the study of something *a priori* given, but as the development of methods for ordering and surveying human experience [2].

If you take "something *a priori* given" to be a quantum world, and you take "methods for ordering and surveying human experience" to be what we can say about nature, then there it is!

Is this correspondence far-fetched? I don't think so. Something *a priori* given that might have been regarded as the object of study of physics before we arrived at quantum mechanics ("our present standpoint") sounds to me like the external world. And is not the "ordering and surveying [of] human experience" just an elaboration of "what we can say about nature"? After all, nature only impinges on us through our experience. And to speak about something, we must order and survey what we know about it.

If that doesn't persuade you, try this:

In our description of nature the purpose is not to disclose the real essence of the phenomena but only to track down, so far as it is possible, relations between the manifold aspects of our experience [3].

If you read this as suggesting that there is a "real essence of the phenomena" that it is not our purpose to describe, then Bohr is saying something quite different from "there is no quantum world." But I would be surprised if he believed in real essences while acknowledging that our description of nature could not disclose them. When he says that physics ("our description of nature") is not about real essences of phenomena, he is saying that quantum mechanics does not mirror an underlying quantum world. And since we have no access to nature except through our experience, what we can say about nature can only consist of an enumeration of the relations between the manifold aspects of that experience.

I doubt that this reading of Bohr will elicit unanimous agreement. But consider this: Half a dozen years after the MIT colloquium, during which I had learned to live with such pale reflections of "there is no quantum world," I was visiting my former postdoctoral adviser, Rudolf Peierls, in Oxford. (I was at his wonderful department in Birmingham for two years in the early 1960s.) Like Weisskopf, Rudi Peierls had also spent time on Blegdamsvej in Copenhagen during the early days of quantum mechanics, and I was curious to get his take on Viki's outburst. So I started telling Peierls the story of my MIT colloquium. I began by reminding him that shortly after Bohr died Aage Petersen had written about his philosophical views in the *Bulletin of the Atomic Scientists* and had attributed to Bohr a certain point of view. More precisely, Petersen had said that Bohr liked to say, "There is no quantum world … ."

When I got to the end of the Petersen quotation, before I could begin to say anything about Viki's reaction at MIT, Rudi beamed at me. "Yes," he said happily, "that's exactly the kind of thing Bohr loved to say!" He did not back down one inch when I reported how Viki had reacted to the same words. He just raised his eyebrows and shrugged his shoulders.

(Once again quotation marks are dictated by the rules of grammar, and their contents are to be viewed in full awareness of the frailty of memory. But I will vouch for the sign of Peierls' response to Petersen's pseudo-quotation: it was distinctly positive.)

So I started using the Petersen pseudo-quote in talks again, always attributing it to the correct source. And I mused on how two of the physicists I admired the most, both well acquainted with Bohr, could have had such diametrically opposite impressions of what Bohr did or did not like to say.

Unfortunately I was never able to get back to Weisskopf with Peierls' reaction to "there is no quantum world. There is only an abstract quantum physical description." But I did conclude that there is no Copenhagen interpretation of quantum mechanics. There is only a range of quantum physical positions. Some are held by Weisskopf's Bohr and some by Peierls' Bohr. There are even positions held by my own Bohr, who, unlike the other two Bohrs, is not constrained by my ever having actually met the man, except for a remote sighting in 1957 from the back row of an enormous auditorium where he spoke for an hour, inaudibly. My Bohr is rather similar to, but considerably more cautious than, Petersen's Bohr. My Bohr would also attach more weight to the word *relations* in reference 3 than I suspect the Bohrs of Petersen, Weisskopf, or Peierls would do.

Setting aside such subtleties, I hope that this column will serve to restore the unfortunately vanishing distinction between

Petersen's Bohr and what we might, for clarity, be better off calling Bohr's Bohr.

References

1. A. Petersen, *Bull. At. Sei.* 19, 8 (1963).
2. N. Bohr, *Essays 1958–1962 on Atomic Physics and Human Knowledge*, Woodbridge, CT: Ox Bow Press (1987), p. 10.
3. N. Bohr, *Atomic Theory and the Description of Nature: Four Essays With an Introductory Survey*, New York: Macmillan, and London: Cambridge University Press, (1934), p. 18. Reprinted in *Niels Bohr, Collected Works*, vol. 6, J. Kalckar (ed.), Amsterdam: North Holland (1985), p. 296.25.

25. Postscript

1. A decade later, the number of Google hits for "no quantum world" has grown from 286 to over 5000. Much of this is probably the result of the growing interest in quantum foundations brought about by the quantum information revolution.

2. When I add "aage petersen" to "no quantum world" the number of hits has grown from 18 to over 1000. The three-fold increase in the fraction that mention Petersen is as likely to indicate the idiosyncrasies of Google searches as the influence of my column.

26

Could Feynman have said this?

Fifteen years ago, I mused in a *Reference Frame* column on how different generations of physicists differed in the degree to which they thought that the interpretation of quantum mechanics remains a serious problem.[1] I declared myself to be among those who feel uncomfortable when asked to articulate what we really think about the quantum theory, adding that "If I were forced to sum up in one sentence what the Copenhagen interpretation says to me, it would be 'Shut up and calculate!'"

In the intervening years, I've come to hold a milder and more nuanced opinion of the Copenhagen view, but that should be the subject of another column. The subject of this one is the habit of misquotation or misattribution that afflicts our profession, a rather different example of which I pointed out a few months ago.[2]

Given my capacity for intellectual development ("inconsistency," in the terminology deployed in the current[3] American political season), it's fortunate that I've now reached an age at which I tend to forget about things I've written more than a few years ago. Indeed, I find it downright irritating when somebody asks me questions about papers I wrote a mere half dozen years ago, naively identifying me with the author of those ancient texts. Until quite recently, I had no memory of ever having written such a childishly brusque dismissal of such an exquisitely subtle point of view, much less of having published it in so widely read a venue.

1 See Chapter 4,
2 Chapter 25.
3 2004.

This amnesia, combined with the evolution in my thinking that had distanced me from my long-forgotten words, may explain why I was initially somewhat puzzled by the slight sensation of discomfort that passed over me when, browsing the e-print archive earlier this year, I read a characterization of Max Born's probability rule as "the favorite ingredient of what has been nicknamed, after Feynman's famous dictum, the shut up and calculate interpretation of quantum mechanics."

I yield to nobody in my admiration for Richard Feynman's aphorisms on the nature of quantum mechanics. Indeed, long ago I published a poem[4] consisting of nothing more than a resetting as verse of a paragraph Feynman had written about his own attitude toward the quantum theory, in his now (but not then) famous article that launched the whole field of quantum computation. I like to think I have devoured everything Feynman ever wrote on the character of quantum mechanics.

But while "shut up and calculate" sounded dimly familiar to me as a characterization of a certain interpretive stance, I couldn't recall where Feynman had written it. Mulling this over, a terrible thought began to dawn on me. Could it be that I myself had once used the phrase? If so, then it would appear that I had picked it up from something by Feynman, forgotten the source, and presented it as my own. Devastating!

It was devastating because I have a horror of writing or uttering any witticism that is not original with me, unless I make it absolutely clear where and (if known) from whom I got it. I don't even like to tell jokes unless I've made them up myself.

(I digress to offer you my favorite:

Question: What is the difference between theoretical physics and
   mathematical physics?
Answer: Theoretical physics is done by physicists who lack the
   necessary skills to do real experiments; mathematical physics

4 *Physics Today*, April 1985, page 47.

is done by mathematicians who lack the necessary skills to do real mathematics. Mathematical physicists tend not to like this joke, but other physicists seem to. Nonphysicists, of course, are entirely immune to its charms.[5])

So with growing trepidation, I searched through my past writings on quantum mechanics. I was dismayed when I came upon my 1989 column, which confirmed my worst fears. Not only had I appropriated without proper attribution a Feynman quote, but it appeared to be a famous one. How humiliating! I was afflicted with visions of knowledgeable *Physics Today* readers shaking their heads 15 years ago at what must have struck them as my shameless attempt to ride to literary glory on the unacknowledged shoulders of Feynman.

So I went to the Web to find the source, hoping I could then salvage my reputation by persuading *Physics Today* to print an addendum or erratum. Google gives more than 130 hits containing both "shut up and calculate" and "feynman." Most of these do not directly link the two, but about a dozen do. Here are a few:

"Shut up and calculate" was a motto of Richard Feynman.

For example, there's Feynman's "shut up and calculate."

My personal philosophy is that of the famous physicist Richard Feynman, who said: "Shut up and calculate."

When asked which interpretation of QM he favored, Feynman replied "Shut up and calculate."

Richard Feynman foreslog ligefrem en "shut up and calculate" fortolkning af kvantemekanikken.

Shut up and calculate – Richard Feynman.

Just to make sure, I also searched for "shut up and calculate" and "mermin." I found only 10 hits, all of them mentioning me in

5 I have more to say about this joke in Chapter 39.

ways that had nothing to do with their use, elsewhere, of "shut up and calculate." So it would have been clear to the world that I had indeed passed off Feynman's words as my own.

Or had I...?

I noticed that not a single one of the websites attributing the phrase to Feynman cited a source or hinted at the circumstances under which he had said it. A ray of hope flickered on: Could I once again have become a victim of the Matthew effect?

The Matthew effect was enunciated by the great sociologist of science, Robert Merton [1]. Merton worked in those innocent days when sociologists were interested only in the behavior of scientists and not in the content of their science. (To be fair to contemporary sociologists of science, I should modify that last phrase to "and not in the manifestations of that behavior in the content of their science.") I first learned of the Matthew effect more than 20 years ago, on the occasion of my victimization at the hands of the *New York Times*.

I learned the name for what the *Times* had done to me when I received a very nice note from P. W. Anderson in which he expressed his regret that the newspaper had given him exclusive credit for a nomenclatural advance that was entirely due to me. "A depressingly typical example of the Matthew effect" was how he characterized the misattribution.[6] When I wrote back asking him what the Matthew effect was, he referred me to Merton.

It was Merton who identified and named the tendency always to assign exclusive scientific credit to the most eminent among all the plausible candidates. At least I hope it was he, though I'm sure Merton, who invented many wonderful jokes himself, would have been delighted if the credit for it turned out to be misattributed to him. Merton named the effect after the Gospel According to Matthew, because there it is written,

---

6 I reported the entire history of this contretemps in the pages of *Physics Today* back in those dark ages (April 1981, p. 46) before there were *Reference Frame* columns.

For unto every one that hath shall be given, and he shall have abundance: but from him that hath not shall be taken away even that which he hath.

– Matthew 25:29.

Could the widespread attribution of my wretched witticism to Feynman be another instance of this same deplorable practice? Had I once again been Matthewed?

Although I didn't say so in my old *Reference Frame* article, what inspired this not so terribly *bon mot* were vivid memories of the responses my conceptual inquiries elicited from my professors—whom I viewed as agents of Copenhagen—when I was first learning quantum mechanics as a graduate student at Harvard, a mere 30 years after the birth of the subject. "You'll never get a PhD if you allow yourself to be distracted by such frivolities," they kept advising me, "so get back to serious business and produce some results." "Shut up," in other words, "and calculate." And so I did, and probably turned out much the better for it. At Harvard, they knew how to administer tough love in those olden days.

This bit of history is relevant to the question of whether Feynman's abundance might have been augmented by a portion of the little that I had. Can you imagine the young Feynman ever having had a similar experience that seared "shut up and calculate" into his tender consciousness? No, of course you can't! Nobody could ever have had the slightest reason to direct the best human calculator that ever was to shut up and calculate.

But perhaps Feynman was offering such advice to others who were searching for a better understanding of the quantum mechanical formalism. I can't believe that. He said that he "always had a great deal of difficulty understanding the world view that quantum mechanics represents," and added, "I still get nervous with it" [2]. Nobody who felt that way would ever respond with "shut up and calculate" to conceptual inquiries from the perplexed.

Well, maybe Feynman, like me, was merely dismissing an interpretive position of others by lampooning it as a "shut up and calculate interpretation." I find this unlikely. For one thing, his strong preference for working things out for himself and, of course, his well-known disdain for philosophy make me doubt that he ever paid much attention to the interpretive positions of others. For another, would one for whom calculation was so effortless and understanding so important be likely to translate anybody's admonition against fruitless speculation into such terms?

In short, I suspect that it is only Feynman's habitual irreverence that has linked him in the minds of many to the phrase "shut up and calculate." Who else among the high and mighty—and Merton has taught us that it is only among the high and mighty that people tend to look—could have said it? Albert Einstein? Don't be silly. Erwin Schrödinger? Of course not. Niels Bohr? Don't make me laugh. None of them besides Feynman could have said it. Does that mean that Feynman said it? No!

Broaden the search to embrace the low and powerless. Among them am I, who hereby put forth the hypothesis that I was the first to use "shut up and calculate" in the context of quantum foundations. I'm not proud of having said it. It's not a beautiful phrase. It's not very clever. It's snide and mindlessly dismissive. But, damn it, if I'm the one who said it first, then that means I did not, even unconsciously, appropriate the words of Richard Feynman and pass them off as my own. So I have nothing to be ashamed of other than having characterized the Copenhagen interpretation in such foolish terms—a lesser offense than unconscious plagiarism, in my moral bookkeeping.

So, dear reader, if you have evidence that Feynman really did say "shut up and calculate," please send it to me. I will not be happy to receive it. I'd rather be a Matthew victim than a plagiarist. But I'd like to know the truth.

References

1. R. K Merton, *Science* 159, 56 (1968).
2. R. P. Feynman, *Int. J. Theor. Phys.* 21, 471 (1982).

26. Postscript

1. My plea to readers for evidence that Feynman characterized the Copenhagen interpretation as "Shut up and calculate" resulted in many emails. With one exception, none of them offered any evidence that Feynman had or had not said it. Each told an entertaining Feynman story that had nothing to do with what I wanted to know.

2. The exception was an email from Richard Feynman's sister, the astrophysicist Joan Feynman:

   I really have no proof at all, one way or another. It may be that my brother had a stomach ache or that somebody kept asking him the same question over and over again until he lost his temper. But if it was Feynman he certainly was angry and annoyed when he said it.

   He brought me up with the Copenhagen interpretation.

3. A decade after the column appeared, the internet is more evenly divided on whether the phrase is due to Feynman or to me. Feynman still seems to be the preferred source, but I'm catching up with him. I'm now fully persuaded that the terrible phrase is mine alone. I'm sorry to have said it. But at least I didn't steal it from Feynman.

27

My life with Einstein

On 25 March 1935, the *Physical Review* received a paper from Albert Einstein, Boris Podolsky, and Nathan Rosen, with the title "Can Quantum-Mechanical Description of Physical Reality Be Considered Complete?" A few days later, on 30 March 1935, I was born. My life with Einstein was off to a promising start.

Some would call it an inauspicious start. Abraham Pais, for example, says in his otherwise admirable biography of Einstein [1] that "the only part of this article that will ultimately survive, I believe, is this last phrase [*No reasonable definition of reality could be expected to permit this*], which so poignantly summarizes Einstein's views on quantum mechanics in his later years." But today, in this centenary of the Einstein annus mirabilis, as the EPR paper and I both turn 70, it is, in fact, the most cited of all Einstein's papers [2]. The debate over its conceptual implications rages hotter than ever, and for the first time, practical (well, for the moment still *gedanken* practical) applications of the EPR effect have emerged in cryptography and in other areas of quantum information processing.

Being only five weeks old, I was unprepared to pay attention to the article that appeared in the *New York Times* on 4 May 1935 under an elaborate set of headlines and subheads:

EINSTEIN ATTACKS QUANTUM THEORY

Scientist and Two Colleagues Find It Is Not 'Complete' Even Though 'Correct'

SEE FULLER ONE POSSIBLE

Believe a Whole Description of 'the Physical Reality' Can Be Provided Eventually.

And I was completely oblivious to the stern rebuke from Einstein himself, published three days later in the *Times*, which declared that "any information upon which the article … is based was given to you without my authority. It is my invariable practice to discuss scientific matters only in the appropriate forum and I deprecate advance publication of any announcement in regard to such matters in the secular press."

Apparently Podolsky had tipped off the *Times* to the article, which did not appear in the sacred press until the 15 May issue of *Physical Review*. It is not clear that Einstein ever forgave him, and I wish I had been old enough to send Podolsky a cross letter myself.

Anyone growing up in America in the 1940s knew that the preeminent genius of our age, and perhaps of any other, lived in Princeton, New Jersey, had a predilection for baggy sweaters, and was always badly in need of a haircut—as far ahead of his time in dress and grooming as he was ahead in science during his 1905 annus mirabilis. Because I had no trouble learning algebra, during the early years of my adolescence I was widely addressed by my peers as "Einstein," as were bright schoolchildren across the nation. Although it was not entirely a term of praise, having definite connotations of "get a life," it never really bothered me. I think this had to do with growing up Jewish and first becoming aware of world news at the time when the Nazi program to murder us was in high gear. I took great pride in the fact that Einstein was Jewish—talk about role models!—and tended to think of him as a kind of distant uncle whom, alas, I had somehow never actually met.

So I was naturally drawn to reading semi-popular books about relativity in junior high school and was under the illusion that I knew quite a bit about it by the time I reached my first physics class in high school. I did not endear myself to the teacher—a retired World War I colonel—when I raised my hand to question his assertion that mass was conserved. To my disappointment he knew even less about relativity than I did. He made it clear that the subject would not and should not ever be mentioned in the class again.

Still under the illusion that I understood special relativity, I submitted a nonsensical derivation of the Lorentz–Fitzgerald contraction to the Westinghouse Science Talent Search, and was invited up to MIT as a regional semifinalist to present my project. At that point I learned that the semifinalists were selected entirely on the basis of a science "aptitude test," and I realized that my derivation was rubbish. So I spent an embarrassing couple of days hobnobbing with MIT faculty and fellow high-school students who did know what they were talking about, and emerged more committed than ever to learning relativity.

This didn't happen until 12 years later when, as a beginning assistant professor at Cornell, I offered a course in relativity to high-school science teachers in a quixotic effort to spare other kids my disappointing encounter with the Colonel. I had to think my way beyond the elaborate formalism I had learned in graduate school to what the subject was really about, putting myself into the shoes of Einstein in 1905 and experiencing for myself the fun he must have had figuring it all out.

After writing a little book based on my course for the high-school teachers [3], I found my life with Einstein expanding in quite unrelated directions. My appointment at Cornell was in a laboratory of atomic and solid-state physics, which made me nervous, since I didn't know anything about solid-state physics. Harvard had relegated the subject to the distant realm of applied physics at the time I was there as a student. So I started to learn it by the time-honored method of writing a book, in indispensable collaboration with my more knowledgeable friend Neil Ashcroft.

And there was Uncle Einstein again, presiding over the birth of the quantum theory of solids in 1907 by pointing out that Max Planck's extraordinary explanation for the high-frequency cutoff in the blackbody spectrum also accounted for the anomalous drop in the heat capacity of solids at low temperatures [4]. A man who had scaled the heights of pure thought to understand that the solution to a problem in electromagnetism lay in the realization that "time was suspect" was also willing to busy himself with a dreary problem in applied physics. This was a marvelous antidote to my Harvard education. That Einstein solved the problem using a method that most of his distinguished colleagues still viewed with deep suspicion made it all the more delicious.

This encounter with Einstein in solid-state physics marked the beginning of a trend. Whenever I moved to a new area of investigation, Einstein greeted me. When I became interested in phase transitions, there he was with a 1910 paper, refining Marian Smoluchowski's observation that density fluctuations at the critical point were the source of critical opalescence [5]. When certain discoveries of my Cornell colleagues Bob Richardson and Dave Lee with their student Doug Osheroff distracted me into the study of superfluids, there was Einstein again with his 1924 discovery that

Satyendra Bose's quantum theory of a gas of identical particles led to a curious condensation phenomenon at low temperatures [6].

Einstein was also an inspiration to me as a writer, having produced some of the finest pieces of scientific exposition I have ever come across. I remember vividly his explanation of why there is a place on the beach between the water and the dry sand where it is easiest to walk. (I have been unable to track down this explanation even with the help of Google—can you, dear reader, supply me with a reference?) What makes the sand firm, Einstein noted, is water that partially covers adjacent grains. Surface tension draws the grains together to minimize the water's interface with the air. If the sand is only a little wet, there are only a few necks of water bridging each pair of nearby grains, and the sand is loose, soft, and difficult to walk on. And if the sand is quite wet there are only a few necks of air, and it is again difficult to walk. But when the sand is neither too wet nor too dry, then there is a lot of air–water interface working to pull nearby grains together and the sand becomes firm and easy to traverse!

To some, "Einstein on the beach" may evoke the Philip Glass opera, but to me it means how to find the easiest place to walk along the shoreline and understanding, thanks to Einstein's lucid explanation, why the method works. I love to walk along the beach. And Einstein is always with me when I do.

His pre-teabag explanation of why tea leaves collect in the center of the bottom of the cup when you stir vigorously, even though the leaves are obviously heavier than the water, is another masterpiece of scientific writing [7].

And when I was in college, during the dark and dreary depths of the McCarthy period, Einstein's forthright and eloquent statements were an inspiration and are not without resonance today [8]:

Every intellectual who is called before one of the committees ought to refuse to testify, i.e., he must be prepared for jail and economic ruin.... This refusal to testify must not be based on the well-known subterfuge of invoking the Fifth Amendment against possible self-incrimination, but on the assertion that it is shameful for a blameless citizen to submit to such an inquisition and that this kind of inquisition violates the spirit of the Constitution. If enough people are ready to take this grave step they will be successful. If not, then the intellectuals of this country deserve nothing better than the slavery which is intended for them.

Einstein could be just as outspoken on scientific matters. In 1928, just three years after the birth of quantum mechanics, he wrote to Erwin Schrödinger [9]:

The Heisenberg–Bohr tranquilizing philosophy—or religion?—is so delicately contrived that, for the time being, it provides a gentle pillow for the true believer from which he cannot very easily be aroused. So let him lie there.

Although my sympathy for this assessment of the Copenhagen interpretation of quantum mechanics waxes and wanes (it is currently waning), this is as eloquent a polemical statement about science by a scientist as I have ever seen.

So I was amazed and delighted when I again felt the presence of Einstein in the pages of the *New York Times* on 24 June 1985. An advertisement announced a sale at Lord and Taylor's department store: "50% off every size—soft-flowered sheets and our own Quantum pillows!" Particularly recommended were "medium support" Quantum pillows, presumably for those who did not rely totally on Niels Bohr and Werner Heisenberg for their peace of mind, but were capable of a little independent thought of their own.

The latest chapter in my life with Einstein has just concluded with the appearance of a new book on special relativity for the general reader [10]. In the course of 37 years I came to realize that in many respects I preferred Einstein's 1905 approach to relativity to my own 1968 effort. But I flatter myself that he would have enjoyed a few of the new wrinkles I bring to the subject, inspired in my expository efforts by his beautiful reading of the tea leaves. That Einstein will be in my future life is assured. For these days, I work in the quantum-information field, where "Einstein locality" is all the rage, and what it means and whether or not quantum mechanics violates it is, in my opinion, still up for grabs. It's strange, though on the whole a good thing, that 2005 has been celebrated as the 100th anniversary of his annus mirabilis, rather than consecrated as the 50th anniversary of his death. Since he'd be 126 years old today, one can't really mourn his absence from the playing fields of quantum information theory, but surely some aspects of that particular game would have given Einstein considerable pleasure, not to mention the unimaginable pleasure we would have had trying out our ideas on him.

References

1. A. Pais, 'Subtle is the Lord ...': The Science and the Life of Albert Einstein, New York: Oxford University Press (1982), p. 456.
2. Here is yet another citation: A. Einstein, B. Podolsky, N. Rosen, Phys. Rev. **47**, 777 (1935).
3. N. D. Mermin, Space and Time in Special Relativity, Prospect Heights, IL: Waveland Press (1968).
4. A. Einstein, Annalen der Physik **22**, 180 (1907).
5. A. Einstein, Annalen der Physik **33**, 1275 (1910).
6. A. Einstein, Sitzungberichte, Preussische Akademie der Wissenschaften **22**, 261 (1924).
7. A. Einstein, Ideas and Opinions, New York: Crown Publishers (1954), p. 250.
8. Reference 7, p. 34.
9. K. Přibram (ed.), Letters on Wave Mechanics, New York: Philosophical Library (1967), p. 31.
10. N. D. Mermin, It's About Time: Understanding Einstein's Relativity, Princeton University Press (2005).

27. Postscript

1. I had promised *Physics Today* that I would write something for the 2005 centenary of Einstein's miraculous year of 1905. The year passed with alarming speed, but I did manage to get something to them in time for the December issue. I had used a similar title in "My Life with [Lev Davidovich] Landau",[1] another wonderful physicist I never met.

2. Nobody sent me a reference for Einstein's essay on where the sand is firmest, but it must exist. If any reader of this book knows where it appeared, please send me a note.

3. Chapters 39–42 give four more *My life with …* essays. All of them are about people I actually did spend some time with.

---

1 Chapter 4 of *Boojums All the Way through*, Cambridge University Press (1990).

28

What has quantum mechanics to do with factoring?

A quantum computer is a digital computer capable of exploiting quantum coherence among the physical two-state systems that store the binary arithmetic information.

To factor an integer is to find its (unique) expression as a product of prime numbers.

The most impressive, most important, and best-known thing a quantum computer can do is to factor with spectacular efficiency the product of two enormous prime numbers. *But what on earth can quantum mechanics have to do with factoring?*

This question bothered me for four years, from the time I heard about the discovery that a quantum computer was spectacularly good at factoring until I finally took the trouble to find out how it was done. The answer, you will be relieved—but, if you're like me, also a little disappointed—to learn, is that quantum mechanics has nothing at all directly to do with factoring. But it does have a lot to do with waves. Many important waves are periodic, so it is not very surprising that quantum mechanics might be useful in efficiently revealing features associated with periodicity.

Quantum mechanics is connected to factoring through periodicity. It turns out, for purely arithmetic reasons having nothing to do with quantum mechanics, that if we have an efficient way to find the period of a periodic function, then, as we shall see below, we can easily factor the product of two enormous prime numbers. And a quantum computer provides an extremely efficient way to find periods.

All of the above is of considerable practical importance, because the great difficulty in factoring such a product—where the two enormous prime numbers are typically each several hundred digits long—is the basis for the security of the most widely used encryption scheme (called RSA [1] encryption) for protecting private information sent over the internet. In 1994 Peter Shor discovered [2] that a quantum computer would be super-efficient at period finding and thereby pose a potential threat to innumerable secrets. Whence the explosion of interest in developing quantum computation. The threat is only potential because no quantum computer capable of anything like serious period finding currently exists.

I suspect the emphasis has been put on factoring rather than period finding because factoring is more famously associated with RSA code breaking, although, as it happens, period finding can be used directly to crack the RSA code, without any need for a detour into factoring. Factoring is also a mathematical concept more familiar to the general public than period finding.

The focus on factoring has led to some spectacular misrepresentations of Shor's algorithm in what Einstein called "the secular press." For example, the New York Times science writer George Johnson says in his book on quantum computation, A Shortcut Through Time,[1] that when the algorithm has done its job, "the solutions—the factors of the number being analyzed—will all be in superposition." Elsewhere he says that a quantum computer can "try out all the possible factors simultaneously, in superposition, then collapse to reveal the answer." Neither of these statements bears the slightest resemblance to what the algorithm actually does.

Such misinformation can give rise to a lot of confusion. The first step to enlightenment is to understand the purely arithmetic connection between factoring and period finding. Then you can forget all about factoring. The link is surprisingly simple, if you're acquainted with modular arithmetic and are willing to accept, as

1 Alfred A. Knopf, 2003.

an empirical fact, that a procedure that *might* give you the factors, if repeated not terribly many times, almost certainly *will* give them.

In modular arithmetic, two integers are said to be equal (or "congruent") modulo a particular integer N (written $\equiv$) if they differ by a multiple of N. Modulo N the infinite set of integers wraps around a circle into the finite set $0, 1, 2, 3, \ldots, N - 1$. Here, for example, are the powers of 3 modulo 5: $3^2 \equiv 4$, because $3^2 = 9 = 5 + 4$; $3^3 \equiv 2$, because $3^3 = 27 = 5 \times 5 + 2$; $3^4 \equiv 1$, because $3^4 = 81 = 16 \times 5 + 1$; and $3^5 = 3$, because $3^5 = 243 = 48 \times 5 + 3$. After that it repeats: $3^6 \equiv 3^2$, $3^7 \equiv 3^3$, $3^8 \equiv 3^4$, and so on; 3^x modulo 5 is a periodic function of x with period 4. Starting at $x = 1$, it produces the sequence 342134213421.... The number 4 has a different period modulo 5. Since $4^2 = 16 = 3 \times 5 + 1$, the sequence produced by 4 is 41414141.... The period is 2.

Why should something as simple as modular arithmetic enable you to do something as hard as factoring the product N of two enormous prime numbers? By itself, it can't. But if you have a really good period-finding machine—and Shor's discovery was that a quantum computer is just such an engine—then there is an easy way to learn the two primes using modular arithmetic. Here is what you do:

Pick an integer a at random. With overwhelming probability, a will not be a multiple of either of those two enormous primes. whose product is N. That being so, it is easy to show that *some* power of a must be equal to 1 modulo N, since modulo N there are only N distinct numbers, $0, 1, 2, \ldots, N - 1$. So if you imagine a list of $N + 1$ different modulo-N powers of a, the list has to contain at least one pair of distinct powers of a with $a^y \equiv a^x$ and $y > x$. This means that $a^y - a^x = a^x(a^{y-x} - 1)$ is a multiple of N. Since a does not contain either prime factor of N, neither does a^x. So if the product of a^x with $a^{y-x} - 1$ is divisible by N, then $a^{y-x} - 1$ must all by itself be a multiple of N, and so $a^{y-x} \equiv 1$.

But if *some* nonzero power of a is equal to 1 modulo N, then there must be a *smallest* such power. If r is the smallest positive integer satisfying $a^r \equiv 1$, then the function $f(x) = a^x$ modulo

N is periodic with period r. So if we have a good period-finding machine—our quantum computer—then we can find r for any a. And with a little bit of luck (we'll return in a moment to just how much luck we need), knowing the period r actually enables us to factor N easily. We need two pieces of luck.

First, suppose we are lucky enough to have picked a random a whose period r is even: $r = 2m$. If $b = a^m$, then $(b - 1)(b + 1) = b^2 - 1 = a^r - 1$, so the product $(b - 1)(b + 1)$ must be a multiple of N. But $b - 1 = a^m - 1$ cannot by itself be a multiple of N, since m is smaller than r, and the period r is the *smallest* power of a with $a^r - 1$ a multiple of N.

Second, suppose we are lucky enough to have picked an a for which $b + 1$ is also not a multiple of N. Then neither $b - 1$ nor $b + 1$ is a multiple of N, although their product is. Since N is the product of two prime numbers, $b - 1$ must be a multiple of one of the prime factors of N, and $b + 1$ a multiple of the other. One factor of N is then the greatest common divisor of $b - 1$ and N, while the other factor is the greatest common divisor of $b + 1$ and N.

Now we are finished. Given any two integers, there is a famous and childishly simple way to find their greatest common divisor, which has been known at least since Euclid. It can be carried out by anybody who can do long division, a skill, to be sure, that I recently learned from the *New York Times* is becoming increasingly rare. With the Euclidean algorithm, an efficient period-finding machine, and the two little bits of luck, we can factor the product of two large prime numbers.

How lucky must we be? Here, and only here, I will hide a rather elaborate argument behind the irritating phrase "it can be shown." It can be shown that if a is picked at random, then the probability of its modulo-N period r being even and $a^{r/2} + 1$ not being divisible by N—the two pieces of good fortune we require—is at least 50%. So if we have a good period-finding machine, it need not work on a^x modulo N for many different random integers a to enable us to find the factors of N. If we pick

20 different random a, then the odds against failure are more than a million to one.

Showing what can be shown is the only hard part of establishing the connection between period finding and factoring. But if you are willing to accept the happy fact that you will surely succeed in under 20 attempts (and you surely will), then you now understand on a practical level how to use a wonderful quantum period-finding machine to factor the product of two large primes.

But before we can get to how the quantum period-finding machine does its magic, another question comes irritatingly, but irresistibly, to mind. What's so hard about finding the period of a periodic function? If I produce a graph of $\sin(kx)$, who needs a quantum computer to figure out the distance $d = 2\pi/k$ over which it starts repeating itself? Aren't repeating patterns easily recognized? Indeed, isn't the recognition of their periodicity the basis of the pleasure we take in them?

Yes indeed, provided the set of numbers whose repetition constitutes the periodic sequence has an easily recognizable structure. But the function whose period you need to learn, if you want to factor the enormous number N, is a^x modulo N. The sequence of integers this specimen churns out as x progresses from 1 up to the (in general huge) period r is virtually indistinguishable from random noise. You can't look for a pattern that repeats itself, because there is no pattern. Nothing you can discern from the sequence gives the slightest hint of when it is likely to start over again. Periodically repeating noise looks locally like—in fact locally *is*—just noise.

One thing that does distinguish the set of numbers that repeats from random noise is that within a period r, no value of a^x modulo N can appear twice. (For r is the *smallest* value for which $a^x \equiv a^{x+r}$.) So one sure way to find the period is just to evaluate a^x modulo N for successive values of x until you finally reproduce the first evaluation. The initial and final values of x are then guaranteed to differ by r.

The problem is that for cryptographic purposes, N is typically a number with 400 or more digits, which typically sets the scale for the number of digits in the period r. So this brute-force approach requires an impossibly large number of evaluations. A more subtle strategy is required. The quantum computational route to period finding has some exquisite subtleties. But that must be the subject of a future column.[2]

References

1. R. L. Rivest, A. Shamir, L. Adleman, *Commun. ACM* **21**(2), 120 (1978).
2. P. W. Shor, in *Proceedings of the 35th Annual Symposium on Foundations of Computer Science*, S. Goldwasser (ed.), IEEE Computer Society Press, Los Alamitos, CA (1994), p. 124; *SIAM J. Comput.* **26**, 1484 (1997).

28. Postscript

This chapter and the next try to explain how a quantum computer, if one could be made, would be able to factor the product of two enormous prime numbers superefficiently. In 2007 I expanded on this greatly in my book[3] on quantum computation. But my explanation is too long and technical to include as an additional chapter in this volume.

2 Chapter 29.
3 *Quantum Computer Science: An Introduction*, Cambridge University Press (2007).

29

Some curious facts about quantum factoring

As explained in Chapter 28, a key to factoring the product $N = pq$ of two prime numbers, each hundreds of digits long, is being able to find the smallest power r for which $a^r \pmod{N} = 1$ for random integers a.[1] Peter Shor's 1994 discovery that a quantum computer would be superefficient at this cryptographically crucial task underlies today's widespread interest in quantum computation. Shor's period-finding algorithm—so called because the function $f(x) = a^x \pmod{N}$ is periodic with period r—illustrates in striking ways the novel basis quantum mechanics provides for computation.

In a quantum computer, a nonnegative integer x less than 2^n is represented by the product state $|x\rangle_n = |x_{n-1}\rangle \ldots |x_0\rangle$ of n two-state systems (called Qbits—if you prefer the vulgar spelling *qubit*, please regard *Qbit* as an abbreviation), where $x_{n-1}, \ldots, x_0$ are the bits (0 or 1) in the binary expansion of x. Quantum-computational architecture executes a function f that takes n-bit integers to m-bit integers, with a linear subroutine $\mathbf{U}_f$ that takes an n-Qbit "input register" initially in the state $|x\rangle_n$ and an m-Qbit "output register" initially in the state $|0\rangle_m$ into the state $|x\rangle_n |f(x)\rangle_m$.

Suppose the initial state of the input register is the superposition of all possible n-Qbit inputs,

$$|\Phi\rangle = \left(1/2^{n/2}\right)\sum_x |x\rangle_n. \tag{1}$$

---

1 Two numbers are equal modulo N if they differ by a multiple of N.

(This can be constructed by starting with each Qbit in the state $|0\rangle$ and applying to each a rotation taking $|0\rangle$ into $\left(1/\sqrt{2}\right)\left(|0\rangle+|1\rangle\right)$.) Since $\mathbf{U}_f$ is linear, if the initial state of input and output registers is $|\Phi\rangle|0\rangle_m$ then their final joint state will be

$$|\Psi\rangle = (1/2^{n/2})\sum_x |x\rangle_n |f(x)\rangle_n. \qquad (2)$$

So it *appears* that just one application of the subroutine evaluates the function f for all the 2^n possible values of x. This trick is called "quantum parallelism."

But appearances can be deceptive. Given a system in a state you know nothing about, there is no way to learn that state. You can only extract information through measurement. If you immediately measure all the Qbits, you acquire a random value x_0 of x and the value f_0 of $f(x_0)$, after which the state becomes $|x_0\rangle_n |f_0\rangle_m$ from which you can learn nothing more. So you could have accomplished as much with an ordinary classical computer by feeding it a random input.

Ah, but suppose you measure only the m Qbits in the output register. When $f(x)$ is $a^x \pmod{N}$, you will find with equal probability any one of the r distinct values f_0 and the input register will be left in a state $|\Psi\rangle$ proportional to

$$|x_0\rangle_n + |x_0 + r\rangle_n + |x_0 + 2r\rangle_n +..., \qquad (3)$$

where x_0 is that random integer less than r at which $a^{x_0} \pmod{N} = f_0$.

This looks promising. By learning only two of the many states $|x\rangle$ appearing in $|\Psi\rangle$, you could learn a multiple of the sought-for period r and be well on the way to learning r itself. With high probability, you could learn such a multiple with two measurements on two sets of Qbits, both in the same state $|\Psi\rangle$. But this route to success is thwarted by the "no-cloning" theorem, which establishes the impossibility of duplicating an unknown quantum state. Something more clever is needed.

The clever move is Fourier analysis, in the form of a spectacularly efficient linear subroutine $\mathbf{U}_{FT}$ that takes the n-Qbit state $|x\rangle_n$ into

$$\mathbf{U}_{FT}\left|x\right\rangle_n = (1/2^{n/2})\sum_y e^{2\pi ixy/2^n}\left|y\right\rangle_n.\qquad(4)$$

Here the eyes of the quantum physicist tend to glaze over. Ah yes, if you go from position to momentum states, then measurement probabilities become sharply peaked at integral multiples j/r of the inverse period, from which the period r itself can be learned. Ho-hum.

But things are not ho-hummish. To lose interest at this stage is to overlook two crucial differences from boring everyday quantum mechanics. First, x has nothing to do with the position of anything, concrete or abstract. It is an arithmetically useful numerical construction out of the states $\left|x_i\right\rangle$ of n independent two-state systems, devoid of physical content: $x = x_0 + 2x_1 + 4x_2 + \ldots + 2^{n-1}x_{n-1}$. Second, "sharply peaked" normally means sharply enough that widths are smaller than the resolution of any detectors. But here one wants an integer r that could be hundreds of digits long. An error of only one part in 10^{10} would get all but a few digits wrong. Such precision lends to the word "sharp" new meaning that no physicist ever dreamed of. All the folklore has to be re-examined.

Re-examination shows that when n Qbits in the state $\mathbf{U}_{FT}\left|\Psi\right\rangle$ are measured, the resulting integer $0 \leq y < 2^n$ has a significant (over 40%) chance of being within ½ of—that is, as close as possible to—an integral multiple of $2^n/r$. So with a little luck, $y/2^n$ is going to be within $1/2^{n+1}$ of j/r for some random integer j. Does this pin down a unique rational number j/r? Suppose there is a second candidate $j'/r' \neq j/r$. The difference between them is $(j'r - jr')/rr'$. Since the candidates are different, the integer $j'r - jr'$ can't be zero, and since the possible periods r and r' are both less than N, the difference between the candidates is at least $1/N^2$. So if $2^n > N^2$, such an integer y does indeed determine a unique rational number j/r.

Of course, j/r determines not r, but r after any factors it has in common with the random integer j have been divided out. So what our quantum computer gives us is a 40% shot at learning a divisor r_0 of r. The odds that j and r have any really big factors in common are small, so chances are that r will be a fairly small multiple of r_0. You can easily check with a classical computer to see if $a^x = 1 \pmod{N}$ when $x = r_0$. If so, $r = r_0$. If not, try $2r_0$, $3r_0$, ..., and with a little luck one of them will work. If you get up to $1000r_0$ without success, then either you were in the unlucky 60%, or the j you got did indeed share a large factor with r. In that case try the whole thing over again. After not enormously many runs, you're quite likely to succeed. And *that* is how Shor's "factoring" algorithm actually works.

But wait a minute! Looking more closely at the crucial subroutine $\mathbf{U}_{FT}$, one finds it to be cunningly constructed out of operations that apply conditional phase shifts $e^{2\pi i\varphi}$ to Qbits, where the values of φ are inverse powers of 2, ranging from $1/2$ to $1/2^n$. Since 2^n must exceed N^2, and N is hundreds of digits long, most of those phase shifts are absurdly tiny—far too tiny for real hardware, with its inevitable imperfections, to produce. All a real quantum computer can execute is an approximate $\mathbf{U}_{FT}$, grotesquely crude on the scale of parts in 10^{300}, the scale on which one needs to learn the period r.

When this dawned on me, I concluded that all the hoopla rested on a silly failure to notice that you can't turn fields on and off for durations you control to parts in 10^{300}. But I was the silly one. I failed to appreciate the exquisite interplay between digital and analog in a quantum computation. Subroutines like $\mathbf{U}_{FT}$ depend on parameters that vary continuously, as in analog computation. But readout through measurement produces an unambiguous sequence of 0s and 1s, as in digital computation. Reading out a thousand Qbits gives you a thousand bits of a definite integer—about 300 digits of the decimal representation of that integer. You learn every one of those 300 digits. The question is whether they are the *right* 300 digits.

Those "huge"[2] uncontrollable phase errors lead to comparable errors in the *probability* that that 300-digit integer will be one of

---

2 Meaning, perhaps, parts in 10^4.

the ones you're looking for. So realistic phase errors might change the probability of getting what you'd like from a little over 40% to a little under 40%. They hardly matter!

A nice illustration of how quantum and classical programming styles differ is provided by the actual calculation of a^x (mod N). Start with a, square it to get a^2 (here I stop writing "(mod N)"—all multiplications are modulo N), square that to get a^4, square that to get a^8, continuing in this way to get all the powers of the form a^{2^j} for $0 < j < n$. Then to get a^x, you simply form the product of all those powers of a^{2^j} for which $x_j = 1$ in the binary expansion of x. But now there is a parting of the quantum and classical ways.

In a classical computer, the two-state systems, called Cbits, used to represent 0 and 1 are cheap and time is precious. If you want to calculate a^x for 2^n different values of x, you use n groups of Cbits to make a table of all the n different a^{2^j} and you look up the various entries going into a^x for each value of x, thereby removing the need to recompute all those squares each time you turn to a new value of x.

But in a quantum computer, Qbits are precious and time is cheap. The multiplication of the appropriate a^{2^j} into a^x is not applied 2^n different times to an input register in each of the states $|x\rangle_n$, but only once, to an input register in the state $|\Phi\rangle$ in which *all* 2^n possible $|x\rangle_n$ are superposed. Each a^{2^j} is used just once. In some terms in the superposition it's multiplied in, and in others it isn't, depending on whether the jth bit of x is 1 or 0. After that single conditional multiplication is carried out, you can square a^{2^j} to get the next power $a^{2^{(j+1)}}$ and store the result in the same group of Qbits that formerly held a^{2^j}, at a huge saving in Qbits.

There is another saving in Qbits to watch out for, because it is misleading. The period r of a^x (mod pq) can easily be shown to be a divisor of $(p-1)(q-1)$. So if $p-1$ and $q-1$ are both powers of 2, as with the primes 3, 5, 17, 257, ..., then so is r. But when r is 2^m, then $2^n/r$ is itself an integer, and measuring the output of the subroutine $\mathbf{U}_{FT}$ can, again easily, be shown to give an integral multiple of $2^n/r$ with probability 1, even when 2^n merely exceeds N but not N^2. Therefore factoring $N = 15 = (2+1)\times(4+1)$ with a 4-Qbit input register does not provide a serious test of the real Shor algorithm. The smallest number that demonstrates the full subtlety of the procedure is 21, which requires an input register of 9 Qbits, big enough to accommodate $(21)^2$.

Another subtlety, only recently pointed out [1], reduces that irritating 60% chance of not getting a divisor of r. When N is the product of two distinct primes, one can (again easily) show that r is not only less than N but less than $\tfrac{1}{2}N$. As a result, besides getting a divisor of r when the measurement yields a y as close as possible to an integral multiple of $2^n/r$, second, third, and even fourth closest also work. This increase in the number of useful outcomes lowers the probability of failure from 60% to 10%, greatly simplifying the subsequent classical detective work.

Features of Shor's "factoring" algorithm like those mentioned here are usually buried in the technical details. By exposing them to the light, I hope to have revealed some of the subtlety and charm of that remarkable procedure.

Reference

1. E. Gerjuoy, *Am. J. Phys.* **73**, 521 (2005).

29. Postscript

1. This companion piece to Chapter 28 is the most technical of a handful of my *Reference Frame* columns that contain some equations.[3] It presupposes familiarity with the basic formalism of quantum mechanics, and with a Fourier transform. *Quantum factoring* has a philosophical flavor, challenging some widely

---

3 Chapters 7, 16, 23, 28, and 29.

held conventional wisdom about quantum computation, and about quantum mechanics.

2. My most important point is that when a set of numbers appears in the name of the symbol ("state") that describes the output of a quantum computer, this does not mean that the computer has actually calculated all those numbers. It is generally impossible to extract from the computer what all those numbers are. The role of the numbers is only to tell you the diverse possibilities available to you for extracting from the computer different kinds of highly limited partial numerical information about that set of numbers.

3. *Physics Today* went along with my quixotic efforts to change the now standard term "qubit" to "Qbit", which is pronounced the same way ("cue-bit"), but has a sensible spelling, and is orthographically parallel to the important companion term "Cbit" (pronounced "sea-bit"). To date the only authority to agree with me in dropping the absurd *u* is Steven Weinberg.[4]

4 *Lectures on Quantum Mechanics*, Cambridge University Press (2013).

30

What's bad about this habit

After we came out of the church, we stood talking for some time together of Bishop Berkeley's ingenious sophistry to prove the nonexistence of matter, and that every thing in the universe is merely ideal. I observed that though we are satisfied his doctrine is not true, it is impossible to refute it. I never shall forget the alacrity with which Johnson answered, striking his foot with mighty force against a large stone, till he rebounded from it – "I refute it thus."
— James Boswell, **The Life of Samuel Johnson**

There is nothing ... more abstract than reality.
— Giorgio Morandi, interview with Edouard Roditi

A bad habit is something you do, without being fully aware of it, that makes life harder than it needs to be. It is a bad habit of physicists to take their most successful abstractions to be real properties of our world. Since the distinction between real and abstract is notoriously problematic, you might wonder what it means to wrongly confer reality on something abstract. I shall illustrate our habit of inappropriately reifying our successful abstractions with several examples.

Perhaps the least controversial examples are provided by quantum mechanics. The quantum state may well be the most powerful abstraction we have ever found.[1] Are quantum states real?

In considering what that question might mean, recall that in the early days Erwin Schrödinger thought that the quantum state

---

1 "Found" is a useful word here, since you can take it to mean "discovered" or "invented," depending on where you stand along the real–abstract axis.

of a particle—in the form of its wavefunction—was as real a field as a classical electromagnetic field is real. He abandoned that view when he recognized that nonspreading wavepackets were a peculiarity of the harmonic oscillator, and that the wavefunction of N particles is a field only in a $3N$-dimensional space.

But that does not prevent advocates of the de Broglie–Bohm "pilot wave" interpretation of quantum mechanics from taking the wavefunction of N particles to be a real field in $3N$-dimensional configuration space. They give that high-dimensional configuration space just as much physical reality as the rest of us ascribe to ordinary three-dimensional space. The reality of the wavefunction is manifest in its ability to control the motion of (real) particles, just as a classical electromagnetic field is able to control the motion of classical charged particles.

Why does reifying the quantum state make life harder than it needs to be? Taking pilot waves seriously can lead you to spend a lot of time calculating, plotting, and proving theorems about the trajectory a (reified) point in configuration space is pushed along by a (reified) wavefunction. The trajectories make no predictions that can't be arrived at using ordinary, trajectory-free quantum mechanics. Their primary purpose is to fortify the view that quantum states are real—a bad habit.

Even for people who don't believe in pilot waves pushing particles, reifying the quantum state can make life harder than it needs to be. It can make them worry about faster-than-light influences in the kinds of experiments first brought to attention by the famous Einstein–Podolsky–Rosen paper. In such experiments a system instantaneously acquires a state as a result of actions confined to the vicinity of a second faraway system that no longer interacts with the first. If the state of the first system is a real property of that system, then something real has clearly been transmitted to the first system from the distant neighborhood of the second at superluminal speed. If the state is merely a useful abstraction, then what, if anything, has been transmitted and where (or to whom) is far more obscure.

Reifying the quantum state also induces people to write books and organize conferences about "the quantum measurement problem" rather than acknowledging, with Werner Heisenberg, that "the discontinuous change in the [quantum state] takes place ... because it is the discontinuous change in our knowledge ... that has its image in the discontinuous change of the [state]."

Admittedly, you can't entirely eliminate the discomfort that gives rise to "quantum nonlocality" and "the measurement problem" by acknowledging that quantum states are not real properties of the systems they describe. But the recognition that quantum states are calculational devices and not real properties of a system forces one to formulate the sources of that discomfort in more nuanced, less sensational terms. Taking that view of quantum states can diminish the motivation for theoretical or experimental searches for a "mechanism" underlying "spooky actions at a distance" or the "collapse of the wavefunction"—searches that make life harder than it needs to be.

Of course, ordinary nonrelativistic quantum mechanics is just a phenomenology—a simplified version of quantum field theory, the most fundamental theory we have about the constituents of the real world. But what is the ontological status of those quantum fields that quantum field theory describes? Does reality consist of a four-dimensional spacetime, at every point of which there is a collection of operators on an infinite-dimensional Hilbert space?

When I was a graduate student learning quantum field theory, I had a friend who was enchanted by the revelation that quantum fields were the real stuff that makes up the world. He reified quantum fields. But I hope you will agree that *you* are not a continuous field of operators on an infinite-dimensional Hilbert space. Nor, for that matter, is the page you are reading or the chair you are sitting in. Quantum fields are useful mathematical tools. They enable us to calculate things.

What kinds of things? Trajectories in spark chambers, nuclear level diagrams, atomic spectra, tunneling rates in superconductors, for example. It's wonderful that the same tool—fields of operators on Hilbert space—works for all those different purposes, but one should not confuse the tool with the reality it helps to describe. Where does the demotion of quantum fields from real things to calculational tools leave the reality of plain old *classical* electromagnetic fields, which represent the kind of reality that Schrödinger initially wanted his wavefunctions to have? When I was an undergraduate learning classical electromagnetism, I was enchanted by the revelation that electromagnetic fields were real. Far from being a clever calculational device for how some charged particles push around other charged particles, they were just as real as the particles themselves, most dramatically in the form of electromagnetic waves, which have energy and momentum of their own and can propagate long after the source that gave rise to them has vanished.

That lovely vision of the reality of the classical electromagnetic field ended when I learned as a graduate student that what Maxwell's equations actually describe are fields of operators on Hilbert space. Those operators are quantum fields, which most people agree are not real but merely spectacularly successful calculational devices. So real classical electromagnetic fields are nothing more (or less) than a simplification in a particular asymptotic regime (the classical limit) of a clever calculational device. In other words, classical electromagnetic fields are another clever calculational device.

What that device enables us to calculate, of course, are classical spacetime trajectories. What about spacetime itself? Is that real? Spacetime is a (3+1)-dimensional mathematical continuum. Even if you are a mathematical Platonist, I would urge you to consider that this continuum is nothing more than an extremely effective way to represent relations between distinct events. And what is an event?

An event is a phenomenon that can usefully be represented as a mathematical point in spacetime. It is thus a phenomenon whose internal spatial and temporal extension *we* deem to be of no relevance to any of the questions that interest *us*. In introducing special relativity in 1905, Einstein, despite his later concerns about physical reality in the quantum theory, was well aware of the abstract character of events. Early in his paper he calls attention to "the inexactness which adheres to the concept of the simultaneity of events at (approximately) the same place, which," he notes, "must be bridged by an abstraction."

So spacetime is an abstract four-dimensional mathematical continuum of points that approximately represent phenomena whose spatial and temporal extension we find it useful or necessary to ignore. The device of spacetime has been so powerful that we often reify that abstract bookkeeping structure, saying that we inhabit a world that *is* such a four- (or, for some of us, ten-) dimensional continuum. The reification of abstract time and space is built into the very languages we speak, making it easy to miss the intellectual sleight of hand. Reifying (classical) electric and magnetic fields is a more recent bad habit, which also came to be taken for granted until it started to unravel with the arrival of quantum electrodynamics, which promoted (or, if you prefer, demoted) the fields to quantum fields—abstract calculational devices.

Why is it a bad habit to reify the spacetime continuum? Well, it can lead one to overlook the nature of some of those events that are abstracted into points. In 1905 Einstein also reminded us that when one says that the train arrives at 7 o'clock, what one means is that "the pointing of the small hand of my watch to 7 and the arrival of the train are simultaneous events." The event used to label the time is associated with the behavior of a macroscopic timekeeping instrument.

Macroscopic clocks have macroscopic spatial extent. Even the best clocks we have—atomic clocks—exploit a transition in a cesium atom, which is huge on the scale of an atomic nucleus, let alone on the scale of the Planck length. And even the size of an atom grossly underestimates the size of an atomic clock, for to make a clock out of cesium atoms you have to tune a cavity into resonance with the transition, which brings us back to the macroscopic level of Einstein's watch.

So when I hear that spacetime becomes a foam at the Planck scale, I don't reach for my gun. (I haven't any.) But I do wonder what that foam has to do with the macroscopic events that spacetime was constructed to represent and the macroscopic means we use to locate events.

Let me put it another way. The raw material of our experience consists of events. Events, by virtue of being directly accessible to our experience, have an unavoidably classical character. Space and time and spacetime are not properties of the world we live in but concepts we have invented to help us organize classical events. Notions like dimension or interval, or curvature or geodesics, are properties not of the world we live in but of the abstract geometric constructions we have invented to help us organize events. As Einstein once again put it, "Space and time are modes by which we think, not conditions under which we live."

In some ways the point may also be easiest to see in quantum physics, where time and space refer ultimately to the time and place at which information is acquired or, if you prefer, at which a measurement is made.

So I'd say that Dr. Johnson had it right when he insisted that what impinges directly upon us is real. The reality of a sore toe is impossible to deny. But the other side of "I refute it thus" is to be suspicious of the reality of those abstractions that help us impose coherence on our immediate perceptions. I doubt that Johnson's valid affirmation of the reality of direct perceptions constituted a refutation of Bishop Berkeley's skepticism about the constructions we find to help us organize those perceptions.

In my youth I had little sympathy for Niels Bohr's philosophical pronouncements. In a review of Bohr's philosophical writings I said that "one wants to shake the author vigorously and demand that he explain himself further or at least try harder to paraphrase some of his earlier formulations." But in my declining years, I've come to realize that buried in those ponderous documents are some real gems: "In our description of nature the purpose is not to disclose the real essence of the phenomena but only to track down, so far as it is possible, relations between the manifold aspects of our experience," and "Physics is to be regarded not so much as the study of something *a priori* given, but rather as the development of methods for ordering and surveying human experience."

I'm suggesting that this characterization of physics by Bohr is as true of classical physics as it is of quantum physics. It's just that in classical physics we were able to persuade ourselves that the abstractions we developed to order and survey our experience were themselves a part of that experience. Quantum mechanics has brought home to us the necessity of separating that irreducibly real experience from the remarkable, beautiful, and highly abstract superstructure we have found to tie it all together.

30. Postscript

1. *Habit* grew out of a talk I gave at a workshop in Phoenix, Arizona the preceding winter. The bland title of the workshop, "The Nature of the Laws of Physics," provoked me to write something controversial. I began by remarking that I was worried that my talk would be viewed as entirely trivial, but I had no idea whether it would be judged trivially right or trivially wrong, which I found encouraging. My distinguished colleagues had little to say after my talk, which I took to indicate little interest.

2. What started me thinking along these lines was not quantum mechanics, but a way of teaching nonscientists about Minkowski spacetime diagrams based on nothing more than a few simple facts they remembered about plane geometry. After several years of refining this pedagogical approach, I found myself starting to think that marks on a piece of paper should not be viewed as representations of spacetime, but that spacetime should be viewed as a way of thinking about—reifying—marks on a piece of paper.

3. I converted my text into a *Reference Frame* column for *Physics Today*, in the hope that I would get more feedback. I did indeed. Four months later[2] *Physics Today* published twelve letters to the editor, nearly all critical, which, with my reply, occupied three times the space of my original column. Here are a few of my remarks about the letters:

Experimentalists have been much more sympathetic than theorists to my views on reification. They seem to be less enchanted by their abstractions.

[It's] important not to succumb to the belief that the correct coordinate system is built into the nature of things, as the Church did … We should choose the one that best suits our purpose … Galileo understood this very well. That's why we talk to this day of Galilean transformations.

The most important culprit caught by special relativity was the reification of time. The ether is interesting as an example of a reification that most of us today agree was unwarranted, even without an explicit general definition of what it means to be real. Clinging to the ether surely made life harder than it had to be.

Evolution has hardwired us to reify. It's important to be aware of the habit, and to be ready to consider questioning even our most successful reifications when they start getting us into serious trouble. "At last it came to me," said Einstein[3], reminiscing about 1905, "that time was suspect." Wow!

Paradoxes and contradictions are a rich source of inspiration. But nature is neither paradoxical nor contradictory. Paradoxes and contradictions reveal our defective understanding of nature. In trying to improve that understanding we [should] keep in mind, among the possible resolutions of a paradox or contradiction, an unacknowledged and inappropriate reification of an abstraction.

---

2 September 2009, pages 10–15.
3 R. S. Shankland, *Am. J. Phys.* **31**, 47–57 (1963).

Identifying the bad intellectual habits that induce us to think there is a quantum measurement problem is just as big a challenge and a lot more fun [than seeking an explanation in a breakdown of quantum mechanics]. I agree that [John S.] Bell … would not have approved of my column. I wish he were here to denounce me. But Bell and I did agree that the Fitzgerald contradiction, contrary to prevailing opinion, was a real physical phenomenon. Although I'm still nervous about defining "real," I haven't changed my mind about the reality of length contraction. No reasonable definition of reality could be expected to omit it.

[Two letter writers] asked whether I believe atoms were real. I polished up my column during a three-month visit to Copenhagen, where I argued at some length with Aage Bohr, Ben Mottelson, and Ole Ulfbeck about their view[4] that all problems in the interpretation of quantum mechanics can be resolved by abandoning "the notion that matter is built of elementary constituents called atoms." I'm still not persuaded that their program makes sense, but I admire their willingness to re-examine even as apparently unassailable a reification as atoms.

4. *Habit* was my final *Reference Frame* column. Chapters 31–33 indicate how my ideas in Chapter 30 developed over the next five years. Chapters 31 and 32 appeared in *Physics Today* after the *Reference Frame* columns had given way to a somewhat shorter *Commentary* section, appearing from time to time at the head of the Letters pages. Chapter 33 is based on a talk I gave in Vienna to celebrate the 50th anniversary of Bell's Theorem in 2014.

4 *Proc. Nat. Acad. Sci. USA* **105**, 17301–17306 (2008).

PART TWO

Shedding Bad Habits

31

Fixing the shifty split

Quantum mechanics is the most useful and powerful theory physicists have ever devised. Yet today, nearly 90 years after its formulation, disagreement about the meaning of the theory is stronger than ever. New interpretations appear every year. None ever disappear.

Probability theory is considerably older than quantum mechanics and has also been plagued from the beginning by questions about its meaning. And quantum mechanics is inherently and famously probabilistic.

For the past decade, Carl Caves, Chris Fuchs, and Rüdiger Schack have been arguing that the confusion at the foundations of quantum mechanics arises out of a confusion, prevalent among physicists, about the nature of probability [1]. They maintain that if probability is properly understood, the notorious quantum paradoxes either vanish or assume less vexing forms.

Most physicists have a frequentist view of probability: Probabilities describe objective properties of ensembles of "identically prepared" systems. Caves, Fuchs, and Schack take a personalist Bayesian view: An agent assigns a probability p to a single event as a measure of her belief that the event will take place [2].

Such an agent is willing to pay less than $\$p$ for a coupon that will pay her \$1 if the event happens, and she is willing to underwrite and sell such a coupon for more than $\$p$. Surprisingly, the standard rules for probability follow from the requirement that an agent should never face certain loss in a single event. (For example, if p exceeded 1, she would pay more than \$1 for a coupon that returned at most \$1; if p were negative, she would pay somebody to take a coupon from her that might cost her another \$1.) Avoiding certain loss is the only constraint on an agent's probability assignments. The probability of an event is not inherent in that event. Different agents, with different beliefs, will in general assign different probabilities to the same event.

The personalist Bayesian view of probability is widely held [3], though not by many physicists. It has profound implications for the meaning of quantum mechanics, which Fuchs and Schack call quantum Bayesianism—QBism for short. Since quantum states determine probabilities, if probabilities are indeed assigned by an agent to express her degree of belief, then the quantum state of a physical system is not inherent in that system but assigned by an agent to encapsulate her beliefs about it. State assignments, like probabilities, are relative to an agent.

QBism immediately disposes of the paradox of "Wigner's friend." The friend makes a measurement in a closed laboratory, notes the outcome, and assigns a state corresponding to that outcome. Wigner, outside the door, doesn't know the outcome and assigns the friend, the apparatus, and the system an entangled state that superposes all possible outcomes. Who is right?

For the QBist, both are right: The friend assigns a state incorporating her experience; Wigner assigns a state incorporating his. Quantum state assignments, like probability assignments, are relative to the agent who makes them.

QBism also eliminates the notorious "measurement problem." Classical probability theory has no measurement problem: An agent unproblematically changes her probability assignments discontinuously when new experiences lead her to change her beliefs. It is just the same for her quantum state assignments. The change, in either case, is not in the physical system the agent is considering. Rather, it is in the probability or quantum state the agent chooses to encapsulate her expectations.

From the beginning, Werner Heisenberg and Rudolf Peierls maintained that quantum states were not objective features of the world, but expressions of our knowledge. John Bell tellingly asked, "Whose knowledge? Knowledge about what?" The QBist makes a small but profound correction: Replace "knowledge" with "belief." Whose belief? The belief of the agent who makes the state assignment, informed by her past experience. Belief about what? About the content of her subsequent experience.

Bell also deplored a "shifty split" that haunts quantum mechanics. The shiftiness applies both to the nature of the split and to where it resides. The split can be between the quantum and the classical, the microscopic and the macroscopic, the reversible and the irreversible, the unspeakable (which requires the quantum formalism for its expression) and the speakable (which can be said in ordinary language). In all cases the boundary is movable in either direction, between ill-defined points. Regardless of what is split from what, all versions of the shifty split are vague and ambiguous.

For the QBist, there is also a split. It is between the world in which an agent lives and her experience of that world. Shiftiness, vagueness, and ambiguity all arise from a failure to realize that like probabilities, like quantum states, like experience itself, the split belongs to an agent. All of them have their own split. What is macroscopic (classical, irreversible, speakable) for Alice can be microscopic (quantum, reversible, unspeakable) for Bob, whenever it is part of her experience but not his. Each split is between an object (the world) and a subject (an agent's irreducible awareness of her or his own experience). Setting aside dreams or hallucinations, I, as agent, have no trouble making such a distinction, and I assume that you don't either. Vagueness and ambiguity only arise if one fails to acknowledge that the splits reside not in the objective world, but at the boundaries between that world and the experiences of the various agents who use quantum mechanics.

Albert Einstein famously asked whether a wavefunction could be collapsed by the observations of a mouse. Bell expanded on that, asking whether the wavefunction of the world awaited the appearance of a physicist with a PhD before collapsing. The QBist answers both questions with "no." A mouse lacks the mental facility to use quantum mechanics to update its state assignments on the basis of its subsequent experience. But these days even an undergraduate can easily learn enough quantum mechanics to do just that.

QBism explains the persistence of the disreputable notion that "consciousness collapses the wavepacket." That is true, but in a banal way. The conscious experience of an agent guides her actions in any number of familiar ways. If she has at least an undergraduate degree in physics, these may include revising, on the basis of new experience, her expectations of future experience embodied in her prior quantum state assignments.

There are glimmerings of QBism in the writings of some of the founders of quantum mechanics. Niels Bohr wrote, "In our description of nature the purpose is not to disclose the real essence of the phenomena but only to track down, so far as it is possible, relations between the manifold aspects of our experience" [4]. Once I thought the crucial word here was "relations"; now I realize it is "experience." Erwin Schrödinger, often philosophically at odds with Bohr, noted, "The scientist subconsciously, almost inadvertently simplifies his problem of understanding Nature by disregarding or cutting out of the picture to be constructed, himself, his own personality, the subject of cognizance" [5]. Here the crucial word is "subject."

I find QBism by far the most interesting game in town. It has not, however, been enthusiastically received by the contemporary quantum-foundations community. Fuchs, in his role as QBism's most fervent advocate, is admired as a provocateur, his more technical work is highly regarded, and he was elected to the leadership of the American Physical Society's topical group on quantum information. But I would say that, with some important exceptions, the general response to QBism has been to shrug it off. I attribute that, in my uncharitable moments, to people having too much fun working on the puzzles that QBism has eliminated.

I write this *Commentary* not to persuade such experts, but to bring QBism to the attention of the much larger community of physicists who have no professional interest in quantum foundations. The message from QBism is this: You needn't feel guilty about never getting nervous about this stuff. You were right not to be bothered. But for the sake of intellectual coherence, you had better re-examine what you wrongly may have thought you understood perfectly well about the nature of probability.

References

1. See, for example, C. A. Fuchs, http://arxiv.org/abs/1003.5209, secs. 1–3.
2. For a short, readable introduction, see R. Jeffrey, *Subjective Probability: The Real Thing*, New York: Cambridge University Press (2004).
3. J. M. Bernardo, A. F. M. Smith, *Bayesian Theory*, New York: Wiley (1994), and the 65 pages of references therein.
4. N. Bohr, *Collected Works*, vol. 6, J. Kalkar (ed.), Amsterdam: North-Holland (1985), p. 296.
5. E. Schrödinger, *Nature and the Greeks, and Science and Humanism*, New York: Cambridge University Press (1996), p. 92.

31. Postscript

1. In 2009 the *Reference Frame* section of *Physics Today* was terminated. Somewhat shorter *Commentary* remarks started to appear from time to time at the beginning of the letters to the editor. Chapters 31 and 32 were in this new format.

2. Several people wrote letters to the editor[1] about *Shifty*. Each letter rejected QBism because the writer had no problem understanding what quantum mechanics was about. The writers all described

1 *Physics Today*, December 2012.

their own understanding, and had nothing more to say for or against QBism. The most striking thing about their various understandings was that they all bore little resemblance to each other. Here are some of my replies and further comments:

- ► Like Barack Obama's view of marriage, my thinking about quantum foundations has evolved. (This was in reply to a complaint that I was contradicting something I had written twenty years earlier.)

- ► The QBist solution to the measurement problem is that a quantum state is not an objective property of the world but a compendium of probabilities constructed by an agent. When an agent updates her state assignment, nothing changes in her external world. The only change is in the agent's expectations for her subsequent experience of that world. That is why there is no "measurement problem".

- ► QBists hold that any agent can apply quantum mechanics to her own external world, from quarks to quasars. Only that agent's personal, directly perceived experience is boxed off from the domain of applicability. That exclusion disagrees with most versions of the many-worlds interpretation.

- ► Alice also cannot apply quantum mechanics directly to the experience of any other agent, Bob, because his immediate personal experience is accessible only to Bob himself. As such it simply does not exist for Alice. But Bob can represent his experience through spoken words, writings, or drawings, all of which do belong to Alice's external world. She can apply quantum mechanics to any of these. The fact that Bob's experience is able to impinge on Alice in these indirect ways clarifies Niels Bohr's oft-repeated emphasis on the importance of being able to state the results of experiments in ordinary language.

▶ The absurdity of quantum nonlocality provides an independent argument that quantum states cannot be objective internal properties of the systems they describe, whether or not one regards probabilities as subjective.

▶ Since writing *Shifty*, I came across another striking anticipation of QBism in a 1931 letter from Erwin Schrödinger to Arnold Sommerfeld: "Quantum mechanics forbids statements about what really exists—statements about the object. It deals only with the object-subject relation. Although this holds, after all, for any description of nature, it appears to hold in a much more radical and far-reaching sense in quantum mechanics."

▶ QBists do not reject the Born rule for calculating probabilities, but they do reject the objective frequentist interpretation of those probabilities held by most physicists. The frequentist view is notoriously circular. It defines probability using such notions as equally probable, unlikely, or identically prepared, none of which make sense without a prior definition of probability. Identically prepared might seem safe from circularity, but two different preparations cannot be strictly identical. It would be more accurate to say they can only differ in unimportant ways. Unimportant for what? For the probabilities of the outcomes.

▶ When I started to learn about subjective probability, I was surprised to discover that most of the books were not in Cornell's physics or mathematics libraries but in the business school library. In the 2012 election, Americans were told that business experience is necessary for being president; I would suggest that it may be even more helpful for understanding quantum mechanics.

▶ One letter writer believes that wavefunctions were collapsing before there were physicists. I wonder if he also believes that probabilities were updating before there were statisticians.

▶ QBism does not give up on a realist interpretation of nature. But it does warn us not to confuse nature with the abstractions[2] we have ingeniously constructed to help any agent deal with the very real impact of nature on his or her own internal experience.

▶ Since there are objective as well as subjective Bayesians, if I had my way, the B in QBism would stand not for Thomas Bayes, but for Bruno de Finetti, who put the crucial point like this:

The abandonment of superstitious beliefs about the existence of Phlogiston, the Cosmic Ether, Absolute Space and Time ..., or Fairies and Witches, was an essential step along the road to scientific thinking. Probability too, if regarded as something endowed with some kind of objective existence, is no less a misleading misconception, an illusory attempt to exteriorize or materialize our actual (*vero*) probabilistic beliefs.[3]

---

2 See Chapter 30.
3 B. de Finetti, *Theory of Probability*, Interscience (1990), preface.

32

What I think about Now

In a *Commentary* in the July 2012 issue of *Physics Today*,[1] I maintain that stubborn problems in the interpretation of quantum mechanics melt away if one takes literally Niels Bohr's dictum that the purpose of science is not to reveal "the real essence of the phenomena" but to find "relations between the manifold aspects of our experience" [1].[2]

This same view of science, that it is a tool each of us uses to organize our own experience, also disposes of the vexing but entirely classical problem of "the Now." Rudolf Carnap states the problem succinctly in his report of a conversation with Albert Einstein:

Einstein said that the problem of the Now worried him seriously. He explained that the experience of the Now means something special for man, something essentially different from the past and the future, but that this important difference does not and cannot occur within physics. That this experience cannot be grasped by science seemed to him a matter of painful but inevitable resignation [2].

The issue here is not, as you might think, that the simultaneity of events in different places depends on frame of reference. Events whose temporal ordering is ambiguous relative to an event that I experience Now are irrelevant, because they are outside the backward and forward light cones of my Now and cannot affect or be

---

1 Chapter 31.
2 See also letters and my further remarks, *Physics Today*, December 2012, p. 8, and July 2013, page 8.

affected by what I experience Now. Whether or not I choose to regard them as contemporary with my Now is entirely a matter of personal convention. The actual issue is that physics seems to have nothing whatever to say about the Now even at a single place, but deals only with relations between one time and another, in spite of the fact that the present moment is immediately evident as such to each and every one of us.

Any person's Now is a special event for that person as it is happening. By an "event" I mean an experience whose duration and location are restricted enough that it can usefully be represented as a point in space and time. My Now is distinguished from other events I have experienced by being the actual current state of affairs. I can distinguish it from earlier events (former Nows) that I merely remember and from later events that I can only imagine. My remembered past terminates in my Now. The status of any particular event as my Now is fleeting, since it fades into a memory with the emergence of subsequent Nows.

Obvious as the human content of the preceding paragraph is, such a Now is absent from the conventional physical description of spacetime. In physics, all the events experienced by a person constitute a time-like curve in spacetime, and there is nothing about any point that gives it a special status as Now. My experience of the Now suggests that my world line ought to terminate in something like a glowing point, signifying my Now. That glow should move in the direction of increasing time, as my world line grows to accommodate more of my experience. There is nothing like this in the conventional physical description of my spacetime trajectory.

The problem of the Now will not be solved by discovering new physics behind that missing glowing point. It is solved by identifying the mistake that leads us to conclude, against all our experience, that there is no place for the Now in our existing physical description of the world.

There are actually two mistakes. The first lies in a deeply ingrained refusal to acknowledge that whenever I use science, it has a subject (me) as well as an object (my external world). It is the well-established habit of each of us to leave ourself—the subject—completely out of the story told by physics.

The second mistake is the promotion of spacetime from a four-dimensional diagram that we each find extremely useful into what Bohr calls a "real essence." My diagram, drawn in any fixed inertial frame, enables me to represent events from my past experience, together with my possible conjectures, deductions, or expectations for events that are not in my past or that escaped my direct attention. By identifying my diagram with an objective reality, I fool myself into regarding the diagram as a four-dimensional arena in which my life is lived. The events we experience are complex, extended entities, and the clocks we use to locate our experiences in time are macroscopic devices. To represent our actual experiences as a collection of mathematical points in a continuous spacetime is a brilliant strategic simplification, but we ought not to confuse a cartoon that concisely attempts to represent our experience with the experience itself.

If I take my Now as the reality it clearly is, and if I recognize that spacetime is an abstract diagram that I use to represent my experience, then the problem of the Now disappears. At any moment I can plot my past experience in my diagram as a continuous time-like curve that terminates in the Now. As my Now recedes into memory it ceases to be the real state of affairs and is replaced in my expanding diagram by subsequent Nows.

The motion of my Now along my trajectory in my diagram reflects the simple fact that as my wristwatch advances, I acquire more experiences to record in the diagram. My Now advances at one second of personal experience for each second that passes on my watch. According to special relativity, this means a second of personal experience for each second of proper time along my trajectory. The connection between my ongoing experience and a geometric feature of my diagram is just my diagrammatic representation of the fact that if asked "What time is it now [Now]?" I look at my watch and report what I read.

There is thus no problem of the Now for any single person. But is there a problem in combining the Nows of many different people? Here we must recognize another obvious fact of human experience. If two people are together at a single event, then if that event happens to be Now for either one of them, then it must be Now for them both. When we are interacting face-to-face, it is simply unimaginable that a live encounter for me could be only a memory for you, or vice versa.

The commonality of my Now and your Now whenever we are together requires that our Nows must coincide at each of two consecutive meetings. That is just what we find whenever any of us move apart and then come back together. But at the slow relative speeds at which we move, the possibly complicating effect of relativistic time dilation on the advances of our individual Nows is utterly negligible compared with the psychological width (many milliseconds) of each of our private Now experiences. Can our Nows coincide when we come back together no matter how rapidly we move back and forth?

The twin "paradox"—the relativistic requirement that personal time keep pace with proper time—assures us that according to physics, the Nows of the traveling and the stay-at-home twin will indeed coincide at their reunion if they have coincided at their separation, even when the departure and return involve speeds comparable to the speed of light. Far from having nothing to say about the Now, physics actually describes it in a way that makes psychological sense, even in a world of many people, all moving about at relativistic speeds.

Erwin Schrödinger had it almost exactly right when he wrote to Arnold Sommerfeld about an "emergency decree" that quantum mechanics "deal only with the object-subject relation. Although this holds, after all, for any description of nature, it evidently holds in a much more radical and far-reaching sense in quantum mechanics" [3]. My only reservation is that although quantum mechanics has indeed forced us—well, at this point only some of us—to recognize that physics is about the object-subject relation, this holds in just as radical and far-reaching a sense in classical physics too.

References

1. N. Bohr, *Atomic Theory and the Description of Nature*, Cambridge University Press (1934), p. 18.
2. R. Carnap, in *The Philosophy of Rudolf Carnap*, P. A. Schilpp (ed.), La Salle, IL: Open Court (1963), p. 37.
3. E. Schrödinger, in *Eine Entdeckung von ganz ausserordentlicher Tragweite: Schrödingers Briefwechsel zur Wellenmechanik und zum Katzenparadoxon*, K. von Meyenn (ed.), Berlin: Springer (2011), p. 490.

32. Postscript

1. I wrote *Now* because the letters responding to *Shifty* (Chapter 31) paid so little attention to what I had said. They were all about the writer's own interpretation of quantum mechanics. I tried in *Now* to make the same point in an entirely classical context, hoping that this would forestall people from responding with their own explanation of what quantum mechanics was really about. This didn't work. There were no responses at all.

2. A distinguished philosopher of science, writing elsewhere about an earlier lecture of mine that evolved into *Now*, derided the attitude that there ought to be a Now on my world-line as "chauvinism of the present moment."[3] He asked why there isn't a problem of the Here. But of course there is. The Here is where I am Now. It's the same problem, solved in the same way.

---

3 Huw Price, *Science* **341**, 960–961 (2013).

33

Why QBism is not the Copenhagen interpretation and what John Bell might have thought of it

Our students learn quantum mechanics the way they learn to ride bicycles (both very valuable accomplishments) without really knowing what they are doing.
— *J. S. Bell, letter to R. E. Peierls, 20/8/1980*

I think we invent concepts, like "particle" or "Professor Peierls," to make the immediate sense of data more intelligible.
— *J. S. Bell, letter to R. E. Peierls, 24/2/1983*

I have the impression as I write this, that a moment ago I heard the bell of the tea trolley. But I am not sure because I was concentrating on what I was writing.... The ideal instantaneous measurements of the textbooks are not precisely realized anywhere anytime, and more or less realized, more or less all the time, more or less everywhere.
— *J. S. Bell, letter to R. E. Peierls, 28/1/1981 [1]*

For the past decade and a half Christopher Fuchs and Rüdiger Schack (originally in collaboration with Carlton Caves) have been developing a new way to think about quantum mechanics. Fuchs and Schack have called it Qbism [2]. Their term originally stood for "quantum Bayesianism." But QBism is a way of thinking about science quite generally, not just quantum physics [3], and it is pertinent even when probabilistic judgments, and therefore

"Bayesianism," play no role at all. I nevertheless retain the term QBism, both to acknowledge the history behind it, and because a secondary meaning remains apt in the broader context: QBism is as big a break with 20th-century ways of thinking about science as cubism was with 19th-century ways of thinking about art.

QBism maintains that my understanding of the world rests entirely on the experiences that the world has induced in me throughout the course of my life. Nothing beyond my personal experience underlies the picture that I have formed of my own external world.[1] This is a statement of empiricism. But it is empiricism taken more seriously than most scientists are willing to do.

To state that my understanding of the world rests on my experience is not to say that my world exists only within my head, as recent popularizations of QBism have wrongly asserted [4]. Among the ingredients from which I construct my picture of my external world is the impact of that world on my experience, when it responds to the actions that I take on it. When I act on my world, I generally have no control over how it acts back on me.

Nor does QBism maintain that each of us is free to construct our own private worlds. Facile charges of solipsism miss the point. My experience of you leads me to hypothesize that you are a being very much like myself, with your own private experience. This is as firm a belief as any I have. I could not function without it. If asked to assign this hypothesis a probability I would choose $p = 1$.[2] Although I have no direct personal access to your private experience, an important component of my own private experience is the impact on me of your efforts to communicate, in speech or writing,

1 For "my," "me," "I," you can read appropriate versions of "each of us"; the singular personal pronoun is less awkward. But unadorned "our," "us," and "we" are dangerously ambiguous. In QBism the first person plural always means each of us individually; it never means all of us collectively, unless this is spelled out. Part of the 90-year confusion at the foundations of quantum mechanics can be attributed to the unacknowledged ambiguity of the first-person plural pronouns and the carelessness with which they are almost always used.
2 I have more to say about $p = 1$ below.

your verbal representations of your experience. Science is a collaborative human effort to find, through our individual actions on the world and our verbal communications with each other, a model for what is common to all of our privately constructed external worlds. Conversations, conferences, research papers, and books are an essential part of the scientific process.

Fuchs himself may be partly responsible for the silly charges of solipsism. One of his favorite slogans about QBism is "Quantum mechanics is a single-user theory" [5], sometimes abbreviated to "Me, me, me!" [6] This invites the *s*-word. I hurled it at him myself the first time I came upon such slogans. Although susceptible to misinterpretation, the slogans are very important reminders that any application of quantum mechanics must ultimately be understood to be undertaken by a particular person[3] to help her make sense of her own particular experience. They do not mean that there is only one user of quantum mechanics. Nor do they require any particular user to exclude from her own experience what she has heard or read about the private experience of others.

Those who reject QBism—currently a large majority of the physicists who know anything about it—reify the common external world we have all negotiated with each other, purging from the story any reference to the origins of our common world in the private experiences we try to share with each other through language. For all practical purposes reification is a sound strategy. It would be hard to live our daily private or professional scientific lives if we insisted on constantly tracing every aspect of our external world back to its sources in our own private personal experience. My reification of the concepts I invent, to make my immediate sense of data more intelligible, is an essential tool of day-to-day living.

But when subtle conceptual issues are at stake, related to certain notoriously murky scientific concepts like quantum states, then we can no longer refuse to acknowledge that our scientific

3 Generally named Alice.

pictures of the world rest on the private experiences of individual scientists. The most famous investigator Vienna has ever produced, who worked just a short walk from the auditorium in which we are meeting, put it concisely: "A world constitution that takes no account of the mental apparatus by which we perceive it is an empty abstraction." This was not said by Ludwig Boltzmann, nor by Erwin Schrödinger, nor by Anton Zeilinger. It was said by Sigmund Freud [7], just down the hill from here, at Berggasse 19. He was writing about religion, but his remark applies equally well to science.

After he returned to Vienna in the early 1960s, Schrödinger repeatedly made much the same point: "The scientist subconsciously, almost inadvertently simplifies his problem of understanding Nature by disregarding or cutting out of the picture to be constructed, himself, his own personality, the subject of cognizance" [8]. In expressing these views in the 1960s he rarely mentions quantum mechanics. Only thirty years earlier, in a letter to Sommerfeld, does he explicitly tie this view to quantum mechanics, and even then, he allows that it applies to science much more broadly: "Quantum mechanics forbids statements about what really exists—statements about the object. It deals only with the object-subject relation. Even though this holds, after all, for any description of nature, it evidently holds in quantum mechanics in a much more radical sense" [9]. We were rather successful excluding the subject from classical physics (but not completely[4]). Quantum physics finally forced (or should have forced) us to think harder about the importance of the object-subject relation.

Niels Bohr, whose views on the meaning of quantum mechanics Schrödinger rejected, also delivered some remarkably QBist-sounding pronouncements, though by "experience" I believe he meant the objective readings of large classical instruments and not the personal experience of a particular user of

---

4 See Chapter 32.

quantum mechanics: "In our description of nature the purpose is not to disclose the real essence of the phenomena but only to track down, so far as it is possible, relations between the manifold aspects of our experience" [10]. Thirty years later he was sounding a similar theme: "Physics is to be regarded not so much as the study of something *a priori* given, but as the development of methods for ordering and surveying human experience" [11]. Bohr and Schrödinger are not the only dissenting pair who might have found common ground in QBism.

The fact that each of us has a view of our world that rests entirely on our private personal experience has little bearing on how we actually use our scientific concepts to deal with the world. But it is central to the philosophical concerns of quantum foundational studies. Failing to recognize the foundational importance of personal experience creates illusory puzzles or paradoxes. At their most pernicious, such puzzles motivate unnecessary efforts to reformulate in more complicated ways—or even to change the observational content of—theories which have been entirely successful for all practical purposes.

This talk is not addressed to those who take (often without acknowledging it) an idealistic or Platonic position in their philosophical meditations on the nature of quantum mechanics. They will never be comfortable with QBism. My talk is intended primarily for the growing minority of philosophically minded physicists who, far from rejecting QBism, are starting to maintain that there is nothing very new in it.[5] I am thinking of those who maintain that QBism is nothing more than the Copenhagen interpretation.

I may be to blame for this misunderstanding. I have used the above quotations from Bohr in several recent essays about QBism, because QBism provides a context in which these quotations finally

---

5 I count this as progress. The four stages of acceptance of a radical new idea are: (1) It's nonsense; (2) It's well known; (3) It's trivial; (4) I thought of it first. I'm encouraged to find that stage (2) is now well underway.

make unambiguous sense. While they made sense for Bohr too, it was not a QBist kind of sense, and I very much doubt that people gave them a QBist reading. Similarly, my quotation from Freud does not mean that QBism should be identified with psychoanalysis, and the three epigraphs from John Bell at the head of this text should not be taken to mean that I believe QBism had already been put forth by Bell in the early 1980s. My quotations from Bell's letters to Peierls are only to suggest that John Bell, who strenuously and elegantly identified what is incoherent in Copenhagen, might not have dismissed QBism as categorically. There are many important ways in which QBism is profoundly different from Copenhagen, and from any other way of thinking about quantum mechanics that I know of. If you are oblivious to these differences, then you have missed the point of QBism.

The primary reason people wrongly identify QBism with Copenhagen is that QBism, like most varieties of Copenhagen, takes the quantum state of a system not to be an objective property of that system but a mathematical tool for thinking about the system.[6] In contrast, in many of the major nonstandard interpretations—many worlds, Bohmian mechanics, and spontaneous collapse theories—the quantum state of a system is very much an objective property of that system.[7] Even people who reject all these heresies and claim to hold standard views of quantum mechanics are often careless about reifying quantum states. Some claim, for example, that quantum states were evolving (and even collapsing) in the early universe, long before anybody existed to assign such states. But the models of the early universe to which we assign quantum states are models that we construct to account

---

6 Heisenberg and Peierls are quite clear about this. Bohr may well have believed it but never spelled it out as explicitly. Landau and Lifshitz, on the other hand, are so determined to eliminate any trace of humanity from the story that I suspect their flavor of Copenhagen might reject the view of quantum states as mathematical tools.
7 In consistent histories, which has a Copenhagen tinge, its quantum state can be a true property of a system, but only relative to a "framework".

for contemporary data. In the absence of such data, we would not have come up with the models. As Rudolf Peierls remarked, "If there is a part of the Universe, or a period in its history, which is not capable of influencing present-day events directly or indirectly, then indeed there would be no sense in applying quantum mechanics to it" [12].

A fundamental difference between QBism and any flavor of Copenhagen is that QBism explicitly introduces each *user* of quantum mechanics into the story, together with the world external to that user. Since every user is different, dividing the world differently into external and internal, every application of quantum mechanics to the world must ultimately refer, if only implicitly, to a particular user. But all versions of Copenhagen take a view of the world that makes no reference to the particular user who is trying to make sense of that world.

Fuchs and Schack prefer the term "agent" to "user". "Agent" serves to emphasize that the user takes actions on her world and experiences the consequences of her actions. I prefer the term "user" to emphasize Fuchs' and Schack's equally important point that science is a user's manual. Its purpose is to help each of us make sense of our private experience induced in us by the world outside of us.

It is crucial to note from the beginning that "user" does not mean a generic body of users. It means a particular individual person, who is making use of science to bring coherence to her own private perceptions. I can be a "user". You can be a "user". But we are not jointly a user, because my internal personal experience is inaccessible to you except insofar as I attempt to represent it to you verbally, and vice versa. Science is about the interface between the experience of any particular person and the subset of the world that is external to that particular user.[8] This is unlike anything in any version of Copenhagen.[9] It is central to the QBist understanding of science.

---

8 See in this regard my remarks above about the dangers of the first person plural.
9 And unlike any other way of thinking about quantum mechanics.

The notion that science is a tool that each of us can apply to our own private body of personal experience is explicitly renounced by the Landau–Lifshitz version of Copenhagen. The opening pages of their *Quantum Mechanics*[10] declare that "It must be most decidedly emphasized that we are here not discussing a process of measurement in which the physicist-observer takes part." They explicitly deny the user any role whatever in the story. To emphasize this they add, "By measurement, in quantum mechanics, we understand any process of interaction between classical and quantum objects, *occurring apart from and independently of any observer*" [my italics]. In the second quotation Landau and Lifshitz have, from a QBist point of view, replaced each different member of the set of possible users by one and the same set of "classical objects". Their insistence on eliminating human users from the story, both individually and collectively, leads them to declare that "It is in principle impossible ... to formulate the basic concepts of quantum mechanics without using classical mechanics." Here they make two big mistakes: they replace the experiences of each user with "classical mechanics," and they confound the diverse experiences of many different users into that single abstract entity.

Bohr seems not as averse as Landau and Lifshitz[11] to letting scientists into the story, but they come in only as proprietors of a single large, *classical* measurement apparatus. All versions of Copenhagen objectify each of the diverse family of users of science into a single common piece of apparatus. Doing this obliterates the fundamental QBist fact that a quantum-mechanical description is always relative to the particular user of quantum mechanics who provides that description. Replacing that user with an apparatus introduces the notoriously ill-defined "shifty split" of the world

10 Translated into English by John Bell, who was therefore intimately acquainted with it.

11 Peierls identifies their positions, referring to "the view of Landau and Lifshitz (and therefore of Bohr)" in his *Physics World* article. He disagrees with all of them, saying that it is incorrect to require the apparatus to obey classical physics.

into quantum and classical that John Bell so elegantly and correctly deplored.

The split Bell complained about is shifty in two respects. Its character is not fixed: it can be the Landau–Lifshitz split between "classical" and "quantum," but sometimes it is a split between "macroscopic" and "microscopic," or between "irreversible" and "reversible." The split is also shifty because its location can freely be moved along the path between whatever poles have been used to characterize it.

There is also a split in QBism, but it is specific to each user. That it shifts from user to user is the full extent to which the split is "shifty." For any particular user there is nothing shifty about it: the split is between that user's directly perceived internal experience, and the external world that that user infers from her experience.

Closely related to its systematic suppression of the user is the central role in Copenhagen of "measurement" and the Copenhagen view of the "outcome" of a measurement. In all versions of Copenhagen a measurement is an interaction between a quantum system and a "measurement apparatus." Depending on the version of Copenhagen, the measurement apparatus could belong to a "classical" domain beyond the scope of quantum mechanics, or it could itself be given a quantum mechanical description. But in any version of Copenhagen the *outcome* of a measurement is some strictly classical information produced by the measurement apparatus as a number on a digital display, or the position of an ordinary pointer, or a number printed on a piece of paper, or a hole punched somewhere along a long tape—something like that. Words like "macroscopic" or "irreversible" are used at this stage to indicate the objective, substantial, non-quantum character of the outcome of a measurement.

In QBism, on the other hand, a measurement can be *any* action taken by *any* user on her external world. The outcome of the measurement is the *experience* the world induces back in that particular user, through its response to her action. The QBist view of measurement includes Copenhagen measurements as a special case in which the action is carried out with the aid of a measurement apparatus and the user's experience consists of her perceiving the display, the pointer, the marks on the paper, or the hole in the tape produced by that apparatus. But a QBist "measurement" is much broader. Users are making measurements more or less all the time more or less everywhere. Every action on her world by every user constitutes a measurement, and her experience of the world's reaction is its outcome. Physics is not limited to the outcomes of "piddling" laboratory tests, as Bell complained about Copenhagen [13].

In contrast to the Copenhagen interpretation (or any other interpretation I am aware of), in QBism the outcome of a measurement is special to the user taking the action—a private internal experience of that user. The user can attempt to communicate that experience verbally to other users, who may hear[12] her words. Other users can also observe her action and, under appropriate conditions, experience aspects of the world's reaction closely related to those experienced by the original user. But in QBism the immediate outcome of a measurement is a private experience of the person taking the measurement action, quite unlike the public, objective, classical outcome of a Copenhagen[13] measurement.

Because outcomes of Copenhagen measurements are "classical," they are *ipso facto* real and objective. Because in QBism an outcome is a personal experience of a user, it is real only for that user, since that user's immediate experience is private, not directly accessible to any other user. Some version of the outcome can enter the experience of another user and become real for him as well, if he has also experienced aspects of the world's response to the user

12 As John Bell may have heard the bell of the tea trolley. Hearing something, of course, is a personal experience.
13 I shall stop adding the phrase "or any other interpretation," but in many cases the reader should supply it.

who took the measurement action, or if she has sent him verbal or written reports of her own experience that he deems reliable.

This is, of course, nothing but the famous story of Wigner and his friend, but in QBism "Wigner's Friend" is transformed from a paradox to a fundamental parable. Until Wigner manages to share in his friend's experience, it makes sense for him to assign her and her apparatus an entangled state in which her possible reports of her experiences (outcomes) are strictly correlated with the corresponding pointer readings (digital displays, etc.) of the apparatus.

Even versions of Copenhagen that do not prohibit mentioning users would draw the line at allowing a user to apply quantum mechanics to another user's reports of her own internal experience. Other users are either ignored entirely (along with *the* user), or they are implicitly regarded as part of "the classical world". But in QBism each user may assign quantum states in superposition to all of her still unrealized potential experiences, including possible future communications from users she has yet to hear from. Asher Peres' famous Copenhagen mantra, "Unperformed experiments have no results," becomes the QBist user's tautology: "Unexperienced experiences are not experienced."

Copenhagen, as expounded by Heisenberg and Peierls, holds that quantum states encapsulate "our knowledge." This has a QBist flavor to it. But it is subject to John Bell's famous objection: Whose knowledge? Knowledge about what?[14] QBism replaces "knowledge" with "belief". Unlike "knowledge," which implies something underlying it that is known, "belief" emphasizes a believer, in this case the user of quantum mechanics. Bell's questions now have simple answers. Whose belief does the quantum state encapsulate? The belief of the person who has made that state assignment. What is the belief about? Her belief is about the implications of her past experience for her subsequent experience.

---

14 Bell used the word "information," not "knowledge," but his objection has the same force with either term.

No version of Copenhagen takes the view that "knowledge" is the state of belief of the particular person who is making use of quantum mechanics to organize her experience. Peierls may come close in a little-known 1980 letter to John Bell [14]: "In my view, a description of the laws of physics consists in giving us a set of correlations between successive observations. By observations I mean ... what our senses can experience. That we have senses and can experience such sensations is an empirical fact, which has not been deduced (and in my opinion cannot be deduced) from current physics." Had Peierls taken care to specify that when he said "we," "us," and "our" he meant each of us, acting and responding as a user of quantum mechanics, this would have been an early statement of QBism. But it seems to me more likely that he was using the first person plural collectively, to mean all of us together, thereby promulgating the Copenhagen confusion that Bell so vividly condemned.

Copenhagen also comes near QBism in the emphasis Bohr always placed on the outcomes of measurements being stated in "ordinary language." I believe he meant by this that measurement outcomes were necessarily "classical." In QBism the outcome of a measurement is the experience the world induces back in the user who acts on the world. "Classical" for any user is limited to her experience.[15] So measurement outcomes in QBism are necessarily classical, in a way that has nothing to do with language. Ordinary language comes into the QBist story in a more crucial way than it comes into the story told by Bohr. Language is the only means by which different users of quantum mechanics can attempt to compare their own private experiences. Though I cannot myself experience your own experience, I can experience your verbal attempts to represent to me what you experience. It is only in this way that we can arrive at a shared understanding of what is common to all

15 Indeed, the term "classical" has no fundamental role to play in the QBist understanding of quantum mechanics. It can be replaced by "experience."

our own experiences of our own external worlds. It is this shared understanding that constitutes the content of science.

Another important difference of QBism, not only from Copenhagen, but from virtually all other ways of looking at science, is the meaning of probability 1 (or 0).[16] In Copenhagen quantum mechanics, an outcome that has probability 1 is enforced by an objective mechanism. This was most succinctly put by Einstein, Podolsky, and Rosen [15], though they were, notoriously, no fans of Copenhagen. Probability-1 judgments, they held, were backed up by "elements of physical reality."

Bohr [16] held that the mistake of EPR lay in an "essential ambiguity" in their phrase "without in any way disturbing". For a QBist, their mistake is much simpler than that: probability-1 assignments, like more general probability-p assignments, are personal expressions of a willingness to place or accept bets, constrained only by the requirement[17] that they should not lead to certain loss in any single event. It is wrong to assert that probability assignments must be backed up by objective facts on the ground, even when $p = 1$. An expectation is assigned probability 1 if it is held as strongly as possible. Probability-1 measures the intensity of a belief: supreme confidence. It does not imply the existence of a deterministic mechanism.

We are all used to the fact that with the advent of quantum mechanics, determinism disappeared from physics. Does it make sense for us to qualify this in a footnote: "except when quantum mechanics assigns probability 1 to an outcome"? Indeed, the point was made over 250 years ago by David Hume in his famous critique of induction [17]. Induction is the principle that if

---

16 A good example to keep in mind is my assignment above of probability 1 to my belief that you have personal experiences of your own that have for you the same immediate character that my experiences have for me.

17 Known as Dutch-book coherence. See the Fuchs–Schack *Rev. Mod. Phys.* article [2] cited earlier. See also Chapter 31.

something happens over and over and over again, we can take its occurrence to be a deterministic law of nature. What basis do we have for believing in induction? Only that it has worked over and over and over again.

That probability-1 assignments are personal judgments, like any other probability assignments, is essential to the coherence of QBism. It has the virtue of undermining the temptation to infer any kind of "nonlocality" in quantum mechanics from the violation of Bell inequalities [18]. Though it is alien to the normal scientific view of probability, it is no stranger or unacceptable than Hume's views of induction.[18] What is indisputable is that the QBist position on probability 1 bears no relation to any version of Copenhagen. Even Peierls, who gets closer to QBism than any of the other Copenhagenists, takes probability 1 to be backed up by underlying indisputable objective facts.

Since we have gathered here to celebrate John Bell, I conclude with a few more comments on the quotations from Bell's little-known[19] correspondence with Peierls at the head of my text.

The first quotation suggests a riddle: Why is quantum mechanics like a bicycle? *Answer*: Because while it is possible to learn how to use either without knowing what you are doing, it is impossible to make sense of either without taking account of what people actually use them for.

The second quotation indicates Bell's willingness to consider concepts, as fundamental as "particle" or the person to whom he is writing his letter, as "inventions" that help him to make better sense of the data that constitute his experience.

18 I would have expected philosophers of science with an interest in quantum mechanics to have had some instructive things to say about this connection, but I'm still waiting.
19 I have had no success finding any of them with Google. For example, there is no point in googling "Bell bicycle." " 'John S. Bell' bicycle" does no better. Even " 'John S. Bell' bicycle quantum" fails to produce anything useful, because there is a brand of bicycle called "Quantum," and Quantum bicycles have bells.

The third reveals a willingness to regard measurements as particular responses of particular people to particular experiences induced in them by their external world. These are all QBist views. Does this mean that John Bell was a QBist? No, of course not—no more than Niels Bohr or Erwin Schrödinger or Rudolf Peierls or Sigmund Freud were QBists. Nobody before Fuchs and Schack has pursued this point of view to its superficially shocking,[20] but logically unavoidable and, ultimately, entirely reasonable conclusions. But what Bell wrote to Peierls, and the way in which he criticized Copenhagen, lead me to doubt that Bell would have rejected QBism as glibly and superficially as most of his contemporary admirers have done.

John Bell and Rudolf Peierls are two of my scientific heroes, both for their remarkable, often iconoclastic ideas, and for the exceptional elegance and precision with which they put them forth. Yet in their earlier correspondence, and in their two short papers in *Physics World* at the end of Bell's life, they disagree about almost everything in quantum foundations. Peierls disliked the term "Copenhagen interpretation" because it wrongly suggested that there were other viable ways of understanding quantum mechanics. Bell clearly felt that Copenhagen was inadequate and downright incoherent. I like to think that they too, like Bohr and Schrödinger, might have found common ground in QBism.

References

1. *Selected Correspondence of Rudolf Peierls*, vol. 2, Sabine Lee (ed.), World Scientific (2009). I have the impression that all three of these quotations are unfamiliar even to those who, like me, have devoured almost everything John Bell ever wrote about quantum foundations.

20 Ninety years after the formulation of quantum mechanics, a resolution of the endless disagreements on the meaning of the theory has to be shocking, to account for why it was not discovered long, long ago.

2. C. A. Fuchs and R. Schack, *Rev. Mod. Phys.* **85**, 1693–1714 (2013).

3. When the QBist view of science is used to solve classical puzzles I have suggested calling it CBism; N. D. Mermin, *Nature* **507**, 421–423, March 27, 2014.

4. H. C. von Baeyer, *Scientific American* **308**, June 2013, 46–51; M. Chalmers, *New Scientist*, May 10, 2014, 32–35. I believe that in both cases these gross distortions were the fault of overly intrusive copy editors and headline writers, who did not understand the manuscripts they were trying to improve.

5. Christopher A. Fuchs, arXiv: 1003.5182.

6. Christopher A. Fuchs, arXiv: 1405.2390, especially pp. 546–549.

7. *The Future of an Illusion* (1927), concluding paragraph.

8. *Nature and the Greeks, Science and Humanism*, Cambridge Univerity Press (1996), p. 92. See also *Mind and Matter* and *My View of the World*.

9. Schrödinger to Sommerfeld, 11 December, 1931, in *Schödingers Briefwechsel zur Wellenmechanik und zum Katzenparadoxon*, Springer Verlag (2011).

10. Niels Bohr, 1929. In *Atomic Theory and the Description of Nature*, Cambridge University Press (1934), p. 18.

11. Niels Bohr, 1961. In *Essays 1958–1962 on Atomic Physics and Human Knowledge*, Woodbridge, CT: Ox Bow Press (1987), p. 10.

12. R. E. Peierls, *Physics World*, January 1991, 19–20.

13. John S. Bell, *Physics World* **3** (8), 3340 (1990).

14. Peierls to Bell, 13/11/1980, *Selected Correspondence of Rudolf Peierls*, vol. 2, Sabine Lee (ed.), World Scientific (2009), p. 807.

15. A. Einstein, B. Podolsky, and N. Rosen, *Phys. Rev.* **47**, 777–780 (1935).

16. N. Bohr, *Phys. Rev.* **48**, 696–702 (1935).

17. David Hume, *An Enquiry Concerning Human Understanding* (1748).

18. C. A. Fuchs, N. D. Mermin, and R. Schack, *Am. J. Phys.* **82**, 749–754 (2014).

33. Postscript

1. This essay did not appear in *Physics Today*. A version will probably appear in the Proceedings of the conference "Quantum

[Un]Speakables II: 50 Years of Bell's Theorem," held at the University of Vienna in June 2014. It is my best attempt to tie together the views of science put forward in Chapter 30, and developed further in Chapters 31 and 32.

2. Even if you disagreed with everything I say here, I hope you enjoyed the quotations from the Bell–Peierls correspondence, which deserve to be better known than they currently are. Nobody at the conference in Vienna had seen or heard any of them. For more about Rudolf Peierls, one of the two heroes of Chapter 33, see Chapter 42.

PART THREE

More from Professor Mozart

34

What's wrong with this book

"**E**nter!" ordered Professor Mozart, as I knocked briskly on his door. Bracing myself against the ambience of stale Havana smoke, I walked in and got right to the point. "W. A., I've got a favor to ask you." He looked up inquisitively. "As you know," I continued, "I have now achieved a modicum of fame as a columnist with *Physics Today*."

Mozart snorted derisively. He is not altogether pleased with my publishing our private conversations. "Because I am a columnist *Physics Today* has decided not to review my recent book of essays. I have thus been unable to bring it to the attention of my literary fans, except by the crude and low-density expedient of inserting advertising transparencies into all my public lectures." His look of derision modulated to outright scorn, but I persisted. "W. A.— Bill—could you write a review of it?"

"Ha!" he offered.

"What does that mean?" I inquired, having learned not to be put off by Mozart's moments of withering contempt. "The book appeared before I invented you, so there's no conflict of interest."

"The problem, my friend, is that I thought your book was dreadful."

"That's not a problem," I explained. "A hostile review will do just as well as a friendly one. The only point is to make my readers aware of the book; they'll judge for themselves. But only one book review has appeared in the entire United States, in an obscure journal of physics pedagogy. Most people don't even know the book exists. Even an unsympathetic review will bring it to their attention and an attack so vicious that it sets them talking would be just what's needed."

"It won't work," he protested. "*Physics Today* doesn't even print Letters to the Editor from nonexistent people, so there's no way they'd do an entire book review."

"Ah, but they might in a *Reference Frame* column," I pointed out, "particularly if the review is negative. So you can pillory me to your heart's content, and still do me a favor. How can you resist!"

A wicked gleam came into his eye. "I'll do it!" he declared, and a week later a manuscript slid under my door which I reproduce *verbatim* below.

Everything's wrong with this book!

Boojums All the Way Through:
Communicating Science in a Prosaic Age
N. David Mermin
Cambridge University Press,
New York, 1990. 309 pp.
ISBN 0-521-38880-5

Reviewed by William A. Mozart

Beginning with its precious and uninformative title and continuing right through to the very last line (where the author characterizes himself as a "soft-hearted esthete") this book offends in more ways than I could have imagined against the canons of scientific writing and simple good taste. The preface is an orgy of name-dropping: The author claims that Brian Josephson and Ken [*sic*] Wilson both once acted in a drama that he penned. He takes pains to let the reader know he is on a first-name basis with a dozen other such luminaries. And he quotes shamelessly from a piece of fan-mail he claims Richard Feynman sent him on his birthday.

The first essay in the book (from which the title derives) reveals that we are in the clutches of a megalomaniac who believes that he is accompanied on his travels by record-breaking extremes of weather. This sorry travelogue recounts his obsessive and, I regret to say, successful campaign to reduce some once noble journals and otherwise sensible colleagues to a level of blithering idiocy. After this fluff we are treated to an assault on the Reagan administration disguised as a commencement address, in which blatantly political diatribes are relieved only by such diversions as a computation, allegedly in a commencement address, mind you, of how big a stack of dollar bills you need to reach the moon.

One has to wade through seven more pieces of a general character before one gets to the "technical" essays. Of the section on quantum mechanics, the less said the better. I will only remark that if you thought the subject was difficult with linear operators, just try doing it all with three-pole switches and red and green lights, and you will never again breathe a mean word about a Pauli matrix.

Having finally gotten this "gedankengadget" off his chest, the author concludes the quantum section with an ill-mannered swipe at Niels Bohr, characterizing the greatest intellectual revolutionary of our age as "extremely cautious," followed by an attempt to smear Sir Karl Popper with the very word "muddle" with which Popper goes after Bohr and Heisenberg.

Finished with his fulminations against two of the towering intellects of our age, the author turns his blunderbuss on relativity, first in the form of "the only Elizabethan drama that is explicitly Lorentz invariant." That this lunatic set of iambic pentameters, blatantly plagiarized from some of the giants of English literature, could ever be characterized as "Lorentz invariant" is a slur on the memory of that great Dutch savant. It is also an abuse of language. If you think not, try "the only Victorian novel that is rotationally invariant." This outrageous piece of doggerel concludes in a two-part "dirge" with the nonsensical title "Such Sorry," set so ineptly in G-minor that it would make a musician wince, to be sung at the preposterous tempo of "*Lento ma non troppo.*"

Not content with thus trashing two distinct art forms, the author next inflicts on the reader a *description* of a computer program, under a title, "The Amazing Many-Colored Relativity Engine," which is most charitably characterized as a thinly disguised attempt to cash in on the success of an old Broadway hit. To make up for his failure to accompany this "User's Manual" with a disk containing the program, the author genially offers to provide one to any reader mailing him a formatted disk in a stamped addressed envelope, secure against actually having to make good on this bluff by specifying that it must have obsolete dimensions. Only in the final subsection does this would-be computer virtuoso reveal that his programming language of choice is Basic, and PC Basic at that! It is doubtless an act of mercy, however unwitting, that he has spared us an Appendix with the source code.

Three derivations of well-known ancient results round out the Relativity section, of which your reviewer only notes that in a "Postscript" to one of them the author cheerily admits that his point was first made in 1910, while in another he requires help from the nonlinear virtuosity of M. J. Feigenbaum to derive a result so elementary that Stephen Hawking felt free to make it the only equation in his celebrated nontechnical book.

But if the Relativity section belabors the obvious, what are we to make of the final section of "Mathematical Musings," which begins with an extended essay on how to compute the base 10 logarithm of 2 to eight-place accuracy with pencil and paper, requiring a mastery of an analytic tool (long division) that most of us haven't seen since the fourth grade! The author's firm declaration that as one carries out this primitive process one will be buoyed "by the ineffable joy of seeing one after another of that amazing stretch of 9's come marching out of the deep interior of log 2" comes close to constituting *prima facie* evidence of a history of substance abuse.

In his preface the author boasts of how he mercilessly wore down a heroic editor who fought valiantly for two years to keep this travesty of analysis out of his pages. Indeed, a sophomoric taste for battling with editors seems to be one of the author's many antisocial habits. He counts among his successes the reversal of George Trigg's scholarly attempt to restore logic to technical spelling, the poisoning of the noble language of topology with baby talk, and the browbeating of an otherwise pure journal into allowing the disease of ordinary punctuation marks to infect the pristine notation of mathematics.

The verse libretto of Purcell's *Dido and Aeneas* was described by a contemporary as "not poetry, but prose gone mad!" Anybody who can characterize the numerical value of 0.7 raised to the power 2^9 as "breathtaking" is writing not about mathematics, but arithmetic gone mad.

The author styles himself a modern-day guardian of scientific prose, holding the gates with his luminous examples against the literary barbarians. In the early 20th century, he informs us, "the art of writing science suffered a grave setback." Somehow the patient struggled on through the century, but in the opinion of your reviewer, this little book may single-handedly finish her off.

34. Postscript

1. This is the only *Reference Frame* column I submitted to *Physics Today* that they would not publish. They said it was in bad taste. They may have been right. But they also did not want to publish a review of my book in any form whatever, genuine or fictitious, friendly or unfriendly, because too much of the book came from their own pages. Presumably they won't review this book either. I include Bill Mozart's hostile review of *Boojums* in this collection because it is part of the Professor Mozart canon.

2. Those unacquainted with my earlier collection *Boojums* should know that it is still in print. The only overlap with *Quark* is that *Boojums* contains the first six of my thirty *Reference Frame*

columns. In 1990 I was just getting underway. I've reproduced those six in *Quark* to make the set of columns here complete.[1]

3. The letter from Richard Feynman whose authenticity Mozart questions has since appeared in Michelle Feynman's collection of her father's correspondence.[2]

4. After rereading Professor Mozart's diatribe, I asked my iPhone calculator (held horizontally to give enough significant figures) to tell me the value of 0.7 squared nine successive times. The result took my breath away! I'd completely forgotten that amazing coincidence. (Once you're in scientific mode you can also just ask it for $(0.7)^{510}$.)

---

1 Toward the end of his review, Professor Mozart says some unfriendly things about Chapters 1 and 6.

2 *Perfectly Reasonable Deviations from the Beaten Track*, Basic Books (2005), 366–368.

35

What's wrong with these stanzas

Longtime readers of *Reference Frame* may remember my opinionated, cigar-smoking friend Professor Mozart, who last appeared in *Physics Today* in August 1999,[1] having dropped in on me after seven years of retirement on a small tobacco plantation in Connecticut. Another eight years having passed, there he was again, at my office door. "W. A.!" I shouted with surprise and delight. "How's the tobacco business?"

"Left it," he growled. "Taken the pledge." And indeed, he gave off no smoky aroma. "Bought a vineyard. Napa Valley. Took up poetry. Thirty-third anniversary of the November revolution of 1974. Third of a century. Time to celebrate with commemorative verses." He sighed. "Brought the project to the attention of the Poet Laureate. Doesn't know enough physics. So," he added in a business-like tone of voice, "I did it myself." And before I could say another word of welcome, he declaimed:

> I have always found it risible
> That "atom" should mean indivisible,
> When deep inside all atoms lie
> Their very tiny nuclei,
> Containing almost all the mass,
> Surrounded by a foggy gas:
> Electrons! We should all be proud
> Of what we know about that cloud,
> Whose properties were once a mystery

1 Chapter 21.

But now belong to science history:
An explanation full and blemish-free
That underlies the whole of chemistry.
And yet the story's even richer.
Within the nucleus we picture
Nucleons—still smaller particles,
About which I could write whole articles.
But let it here suffice to say
The two varieties that they
Possess. (1) Protons with a charge
Of electricity just large
Enough to hold electrons in
Those clouds in which they whirl and spin.
(2) Neutrons: uncharged partners of
The protons, held to them as love
Binds lovers in their warm embrace,
Though bound by mesons in this case.
What could be to God's greater glory?
And yet there's much more to the story!
The nucleons themselves have pieces
Described in many a doctor's thesis.
The Standard Model is the name
Of this subnucleonic game.
It underlies the interplay
Of everything. And so, I say:
Enough of idle talk and twaddle –
Let's celebrate the Standard Model!

At this he took a deep bow. "Very nice, Bill," I said politely, "but don't you think 18 unrelieved couplets get a bit bumpy?"

"The uniformity is relieved by the random variation of 8 feminine (*twaddle-model, glory-story*, ...) and 10 masculine (*name-game, charge-large*, ...) rhymes. And if you hadn't interrupted me so rudely, you would have found out that the crude simplicity of this prologue offsets the splendor of the hymn to the Standard Model that follows in 12 Onegin stanzas! It's the poetic equivalent of the switch from black and white to color, when Dorothy walks out the door into the Land of Oz!"

"Onegin stanzas?" I inquired.

Muttering "*Nekulturniy*—you're not in Kansas anymore" under his breath, Mozart sighed, and then explained patronizingly, "Pushkin's great verse novel, *Eugene Onegin*, is composed entirely of 14-line stanzas in iambic tetrameter, with the rhyme scheme *aBaBccDDeFFeGG* with *a, c, e* feminine and *B, D, F, G* masculine. People," he added more brightly, "have been writing physics poems at least since James Clerk Maxwell. But nobody—not John Updike, not even Douglas Hofstadter—has ever attempted to match the intricacy and harmony of fundamental physics to that exquisitely subtle and demanding form."

"How can anybody who rhymes 'richer' with 'picture', not to mention 'blemish-free' with 'chemistry', hope to imitate Pushkin?" I inquired, unable to disguise my irritation.

"Hush!" he ordered. "Listen!" And *espressivo molto*, he began to recite.

The Standard Model

(1)

The up and down quarks are the units
That make both nucleons behave
In just the manner that they do. It's
Extremely simple. Note that they've
Both got inside them little pieces.
The nucleons are like valises,
Containing quarks. We now agree
A nucleon is made of three
Internal quarks. Quite elementary:
The protons are two ups, one down,
All bound together like a crown.
In Nature's book another entry:
The neutrons are two downs, one up,
Together, just as in a cup.

(2)
The protons all have just one unit
Of charge, electrical. Although
The neutrons lack charge, yes they do! (It
Permits them easily to flow
Through crystals.) But this neat arrangement
Requires some genuine estrangement
From common notions. Thus, the charge
Of up quarks isn't all that large:
Only two-thirds the normal portion
That one expects a charge to be.
What's more, the down quark, cunningly,
Has just *one*-third. And here's a caution:
Up and down charges, that combine,
Are furnished with a different sign.

(3)
So up-down-down has full charge zero
While up-up-down has full charge one.
It doesn't take a superhero
To realize the job's now done.
Our tripling has been quite delightful:
The proton's charge is just its rightful
Amount. The other way quarks fall,
The neutron gets no charge at all.
Next, you may ask what is the tether
That holds these trios in one place
Instead of wandering through space:
What makes three quarks remain together?
The gluons! Gluons do the job
Of making quarks a single blob.

(4)

The way in which the gluons tie in
A group of quarks to just one place
Deserves a mention very high in
Examples of a strong embrace.
For if you try to tear asunder
Three quarks you make a dreadful blunder:
You'll find the harder that you try,
The stronger bond the gluons tie.
For gluons bear the strict assignment
Not to let single quarks appear.
I ought to make that very clear.
This property is called confinement.
However much you try to free
A single quark, it cannot be.

(5)

Each nucleon's a three-quark triple
One up, two downs; two ups, one down.
But there's another little stipul-
Ation that has acquired renown.
All quarks have color. This is knowledge,
Although you'll learn in any college
It's not the kind that you can see
But just a form of poetry.
The colors don't come in profusion.
One quark is red, one green, one blue.
That's all there is. I'm telling you
To ward off possible confusion.
And here's a final piece of news:
The anti-quarks have anti-hues.

(6)

The anti-red-quark, for example,
Has to be colored anti-red.
Likewise, I'm sure it wouldn't trample
The preconceptions in your head
To learn the hue of anti-blue-quarks
Is anti-blue. And like all true quarks
The anti-green-quark has to mean
A quark that's colored anti-green.
Now here's a rule that's quite delicious:
The red and green and blue make white.
The color's disappeared, all right.
And though it may not sound propitious
When color, anti-color play
All of the color goes away.

(7)

All particles are colored purely
Neutral if they're directly seen.
In nucleons the three quarks surely
Must then be red, and blue, and green.
Another way that's somewhat duller
To make up something lacking color
Directly puts together two
Paired quarks of hue and anti-hue.
In just this way one makes a meson,
Like pion, eta, kaon, rho
And other ones you might not know
Unless you have a special reason.
Only the parts have color. Yes,
The things we see are colorless.

(8)

Now up and down are not the only
Varieties that quarks possess.
In case you worried they'd be lonely,
You needn't fear. For I confess
There are two other generations
That furnish quarkish explanations
Of why there is so big a zoo.
Each generation comes with two
Fine quarkish partners. One pair's charmed
And strange. The other's bottom, top.
This makes you wonder, will it stop?
No problem. Do not be alarmed!
There's nothing else we need to fix.
The quarkish flavors stop at six.

(9)

Now all of us have often spoken
About electrons, tiny things.
They too have partners (I'm not jokin'!)
That swiftly fly as though on wings.
Neutrinos have the smallest masses
Of all the fundamental classes
Of particles. As I recall
They've hardly any mass at all.
Through endless ranks of rocks and boulder
At speeds approaching that of light
They pass unhindered in their flight.
There's almost nothing that can shoulder
Those swift neutrinos off their path.
We live in a neutrino bath!

(10)

Here too things come in generations.
A heavy 'lectron called the mu
Is one of God's more strange creations
And has its own neutrino, too.
A third neutrino fills the picture.
Following a familiar stricture,
Which surely isn't broken now,
Its partner's the still heavier tau.
Electron, mu, and tau along with
All their neutrinos, commonly
Are known as leptons. These make three
New generations that belong with
The three quark pairs. The same format!
Well might one ask, "Who ordered that?"

(11)

A flavor-changing transmutation
Is called a weak decay. And there's
Accompanying it the swift creation
Of lepton–antilepton pairs.
The down-quark–up-quark weak transition –
"Beta decay" (from old tradition) –
Creates electrons, moving fast,
And antineutrinos in the blast.
This is the way that neutrons trouble you,
Turning to protons with the aid
Of two assistants, that are made
Of new gauge fields. Called Z and W
They're massive cousins to a sprite:
The massless photon, speck of light.

(12)

Behold the Higgs! The most elusive
Of particles, I do insist.
Not yet observed, we lack conclusive
Data to show that they exist.
Without the Higgs there'd be no masses
Creating some grotesque impasses.
This means within the Standard tale
The Higgs's the crucial Holy Grail.
Please do not doubt the Higgs' reality.
We're hoping very soon to learn
From evidence produced at CERN
That it exists, with clear finality.
So don't forget: *The Higgs's the thing*
That makes the Standard Model king.

Mozart uttered the last line and a half in italics, then stood quivering with his eyes shut. For me, the lines resonated with something. *The Cray's the thing that puts our computations on the wing?* No. Or perhaps the more esoteric, *The splay's the thing that gives nematic liquid crystals zing?* Not really. Oh, I know —

But just as it was coming to me, Bill Mozart emerged from his reverie, gave me a little wave of his hand, and disappeared down the corridor. He left behind him 18 couplets and 12 Onegin stanzas, handwritten in his characteristically florid strokes. I have reproduced them here, without attempting to correct their imperfections, for the edification of the physics community on this, the one-third centennial of the November revolution.

35. Postscript

1. Because it was too long for the *Reference Frame* format, *Stanzas* appeared in *Physics Today* as an ordinary article. It belongs in this collection because it contains the final performance of Professor Mozart, and demonstrates his deep love of particle physics. I needed his help making sure that readers knew what

Onegin stanzas were. That meant they also had to know about masculine and feminine rhymes. It would have been patronizing for me to lecture the reader about these matters. But if Professor Mozart was the poet, then *he* could explain them to *me*. A second benefit of making Mozart the poet was that I could state outright that some of the rhymes were outrageous.

2. I was asked to write a poem about the standard model by an Ithaca group called *Music's Recreation*. They needed verses to accompany *Carnival of the Subatomic Particles*, which Mark G. Simon had composed for flute, clarinet, violin, cello, and piano to celebrate the 30th anniversary of the first beam injection at Cornell's [Robert Rathbun] Wilson Synchrotron Laboratory. *Carnival* was performed to great acclaim in Ithaca on April 1, 2007. For the version in *Physics Today* I changed the anniversary to the 1/3 centenary of the discovery of the charmed quark, an event widely known to physicists of the time as the "November revolution."

3. Writing these verses turned out to be a bigger project than I had anticipated. Onegin stanzas are a major challenge. Another was being sure I wasn't saying anything wrong about the standard model. It's not my branch of physics.

4. The discovery of the Higgs was announced at CERN on July 4, 2012, fulfilling the hope Bill Mozart expressed five years earlier, in his final verse. He finds it ironic that a discovery that would have been made years earlier in Texas, had Congress not canceled the Superconducting Super Collider, was finally made in Europe and announced on America's Independence Day.

PART FOUR

More to be Said

36

The complete diary of a Nobel guest

Friday, December 6. We arrive in Stockholm at 8 am. I change a heap of dollars to kronor. Light rain falls. The expected limo does not appear. Our luggage is stuffed with white-tie costume, dark suits, evening gowns, newly acquired white shirts and ties, and provisions for winter weather of all possible kinds. A minivan takes six of us and a prodigious number of bags to the Grand Hotel. We hand over a heap of kronor to the driver. Our room is not ready. I explore the public facilities. Today's *Herald Tribune* is posted in the men's room. Madeleine Albright is the new Secretary of State. Now the room is ready. Conserving my remaining kronor I hand the bellhop a $5 bill.

The Grand Hotel is selectively grand. The bathroom is magnificent but there are not enough closets for two Nobel guests, no bureaus whatever, and the desk drawers are stuffed with phone books. We set up camp for the week making ingenious use of all available surfaces. We go for a walk. We forgot to bring an umbrella, but never mind. The Grand Hotel provides. We cross a bridge into Gamla Stan, the old town. Cobblestones glisten. The buildings are lovely in the pale light. We return to the Grand for a brief nap. I discover I left important pills home. The Grand Hotel rises to the occasion. I am instantly connected with a soothing doctor who asks appropriate questions, consults appropriate texts, and sends me to the nearby Lion pharmacy. Pills are sold in units of 100. I have to acquire 5 times the needed number. An enormous fee for the soothing doctor is added in. I put it on the Visa card.

Night comes at 3 pm. We dine at an unpretentious little spot where I hand over my remaining kronor. We take an after-dinner walk to an ATM machine and get another heap. Back at the Grand we find Dave Lee and Bob Richardson newly arrived from Gothenburg. They wear tiny gold lapel pins throughout the week so reporters and autograph collectors can tell them from their guests. They have already been celebrating for several days in Gothenburg. Bob has a bad cold. Both are very happy and a little high. I get a perfect 8 hours sleep, but the first night is always easy.

Saturday, December 7. The breakfast buffet at the Grand is phenomenal, and attended by many old friends from the glory days of superfluid helium-3. A sunny day! Black stretch limos—one per laureate—take the physics and chemistry winners to their lectures. Their guests follow in tour buses. The lecture hall is surprisingly small. The front rows are reserved for the Nobel guests. Physics (morning) takes precedence over Chemistry (afternoon) and the oldest (Lee) speaks first, the youngest (Doug Osheroff, the graduate student, now 51), last. The atmosphere (and room) are surprisingly reminiscent of APS meetings except that most wear dark business suits ("informal" attire) and both the overhead and slide projectors are remarkably recalcitrant. The physics lectures are highly evocative of youthful memories from the early 1970s. Bliss was it that dawn to be alive. The lunch is good and large. We resolved before departure not to worry about weight, though Thanksgiving had already taken its toll. The chemistry talks are also fun since buckyballs are really physics. Or is superfluid helium-3 really chemistry? Both of this year's prizes are for something discovered accidentally while looking for something else. It makes for good lectures.

Back to the Grand in the dark. I get a report on the literature prize lecture by Wislawa Szymborska from those who cut chemistry to attend. Who would have expected parallel sessions? We dine at a small unpretentious place frequented by the locals. I hand over a big part of the replenished hoard. I am awake half the night.

Sunday, December 8th. The laureates are busy all day with preparation for a CNN show and a press conference. Their guests are free, the weather is dry, the city is beautiful. The art museum is amusing and has three great Rembrandts. I collapse at 2, awakening from my nap in darkness at 3. The bus takes us to an informal reception. I wear the blue suit acquired last year for my son's wedding. There is sensational smoked salmon and other unidentifiable delicious fishy things. I spend the party ingesting delicacies in the wrong room but realize my error in time to make the bus back to the Grand. There is a great reunion of the old helium-3 crowd at dinner. The laureates can't make it, having a mandatory "Informal dinner (dark business suit)" at the Royal Swedish Academy of Sciences. We go for the buffet. First cold fish, then cold meat, then hot fish, then hot meat. The cold fish is so good I break the rule: first cold fish, then more cold fish, then stuffed. The bill is enormous. We pool all our cash and have barely enough left to ransom our coats from the smiling attendant, at 10 kronor per. I refill the wallet at the nearest ATM machine, and then off to bed. I am awake most of the night.

Monday, December 9th. We skip the bus tour of Stockholm to attend the economics lecture. Our guest status is again good for front row seats. We hear about the theory of auctions. There are integrals and derivatives. It's like physics except physics works. The American laureates have lunch with the Ambassador. Poor laureates. The guests have now learned not to eat lunch between the breakfast buffet and the late afternoon reception. Today's dwarfs yesterday's: the Apotheosis of Informal Dress. Fortunately I am now wearing my black suit, which blends right in. Two enormous rooms flank one gigantic one. Food and drink are impossible to resist. I succumb to earthly delights until the time to depart for dinner in a gorgeous baroque clubroom with the Swedish Cornell

Alumni. The laureates are also allowed to attend. Their stomach capacity is even more prodigious than ours, but they've also been working harder. I sleep all night.

Tuesday, December 10th. The big day. The women have their hair set in the morning and are confined to quarters thereafter. The laureates are off at a mandatory rehearsal (casual). I take a long walk along the water to City Hall to check it out. There are many delivery vans and mysterious stacks of wood. Strange waterfront sculptures. I return to the Grand by way of the Café Access, discovered before leaving Ithaca in a Web exploration of Stockholm. It is the only bargain in Sweden: 20 kronor for half an hour on the internet. (A bratwurst from a street peddler costs 25.) The system manager doubles as waiter. My treatment shifts from brusque to cordial when he learns he is logging in a Nobel guest. I send email greetings to the kids, and a bulletin to the Cornell Physics Department.

Back at the Grand I don the required white tie and tails. All in the Cornell delegation—even the laureates, who unaccountably have been given an unprogrammed moment—meet in our finery in the huge 2nd floor elevator lobby with half a dozen cameras. Bulbs flash. There is general hilarity. Then off we go to the Concert Hall, laureates in their limos, guests in their "Nobelbil" (the old buses with new signs). Police hold back throngs as we ascend the majestic steps. The throngs are awaiting their Queen and not interested in us. We climb many more internal steps and deposit our coats in a gigantic cloakroom. Advance warnings of inadequate toilet facilities prove to be a scurrilous rumor. As guests of the oldest laureate in the number-one field we get front-row center-balcony seats. The view is superb. I never knew what "glitter" meant before. The most opulent production of *Die Fledermaus* falls far short of this tableau. The Concert Hall glows. We spot the laureate families in the front-row orchestra. Members of the Swedish Academy are seated on the stage. A full orchestra and singers are ready in a great balcony above it. Nine empty bright red chairs are stage front left; three empty dark chairs, stage front right. I admire my fellow members of the audience. There are jewels and medals in all directions. I check the abundant flowers festooning the edges of the balcony. Real, of course. Is this a dream?

There is a flourish of trumpets promptly at 4 followed by the entrance of the laureates to an obscure Mozart march. They are ordered by field and within field by age and therefore led by the Cornell delegation. I burst unexpectedly into tears of joy. It is remarkably like a wedding. The laureates seat themselves. Royal music follows. All rise as the King, Queen, and Royal Aunt enter. The King sits down opposite the laureates a split second before the Queen. The Royal Aunt and everybody else follow. The hundredth anniversary of the death of Nobel is the occasion for a long biographical speech. Swedish is heard for first time in the festivities, but pamphlets with English text are kindly provided. I learn that Nobel invented the word "dynamite" as well as the stuff. "Dynamite" surely describes the present scene.

The orchestra contributes the overture from a Swedish opera, a Finnish soprano gives us a recitative and aria from *Idomeneo*, and the orchestra concludes with a little Mendelssohn. The awards begin, physics first, of course. We are still in Swedish, but English libretti are provided. I proudly recognize a sentence of my own from an old nominating letter. The libretto has blanks where italic v's should be: the Physics and Chemistry citations talk of disco eries, the Economics citation of incenti es, and a poem of Szymborska is entitled Disco ery. This is the only discernible imperfection in the whole show, possibly put there deliberately lest the gods themselves become jealous.

One by one the laureates are called forth to engage in complex hand maneuvers with the King. An attendant hands the King a large citation with a box containing the medal resting on top. The King extends same to the laureate with his left hand. The laureate grasps the other side of the citation with his left hand. With the left hands thus indirectly linked, the right hands make direct contact in a handshake beneath the citation, shielding the pressed flesh from direct public view. After the shake the King lets go and the laureate is left holding the citation alone. The medal, though clearly not attached, does not fall off. There are trumpet flourishes while the laureate bows to the King, the members of Academy, and the audience. All burst into warm applause during the laureate's normal—you don't have to walk backwards—return to his red chair.

After physics and chemistry come Sibelius, then Medicine, Grieg, then Literature. Szymborska bows in the wrong directions, earning a prolonged ovation. Don Giovanni then sings seductively from on high accompanied by Leporello playing a real mandolin, before switching into an exceptionally spirited "Fin ch'han dal vino." Don Giovanni does not shatter his glass on the dignitary-loaded floor below him, and the awards conclude with Economics. William Vickrey, who died a few days after the announcement, is represented by an old friend. All stand for the National Anthem as the King, Queen, and Royal Aunt leave the stage. The audience reluctantly drifts toward the cloakrooms. Photographers rush onto the stage. They all want Szymborska. Family members of the laureates join them on stage, everybody shaking hands with everybody else. It is eerily evocative of the post-debate rituals in the recent[1] American presidential election.

We collect our coats and make our way down many stairs to the street, past admiring throngs of Stockholmers, to a great line of red municipal buses, each become a Nobelbil for the occasion. Each Nobelbil is filled to capacity. Student riders kindly offer seats to the ladies. I glance around at fellow white-tie straphangers. Was ever a stranger sight seen in the Arctic night? Off we go to the Stadhus, transformed from the City Hall I visited that morning into an enchanted palace. The logs have become great bonfires on the waterfront. Various sinuous pieces of sculpture noted by the water in the morning are also now ablaze. From the entrance to the courtyard to

1 1996.

the top of the great stairway in the opposite corner is a double row of candle-carrying Cub Scouts between whom we pass. I smile at the scouts, I frown at the scouts, I salute the scouts. No eye contact is possible. This is serious business. At the end of this corridor of fire we emerge in an enormous hall—the grandest cloakroom of them all. We deposit our coats, emerging like butterflies in full regalia.

How do you seat 1300 dinner guests? It's easy. You hand each of them a 71-page book listing the table (A for the head table, holding 88, 1–65 for the other tables, variously holding 10, 20, or 30) and the seat number at that table for each diner. The diners are listed alphabetically except for the King, Queen, and Royal Aunt, who appear in that order at the top of page 1. You attach to the back of the book an enormous fold-out map of the banquet hall, showing every table and every position at each. Finally you include a separate blown-up map of each table, giving the names and positions. This enables all to double check what they found on the alphabetic list, as well as giving them the names of the neighbors they encounter when they have made their way—as they do with surprising ease—to their places.

When the 1200+ occupants of Tables 1–65 are all in place on the vast floor of this roofed-in Venetian courtyard, the orchestra launches into a *marche triomphale* and the diners rise, as from high up in a far corner of the enormous enclosure a slow procession emerges onto a balcony, traverses the entire length of a side wall, and descends a monumental staircase at the far wall, to take their places at the head table (A) that stretches back nearly the full length of the floor. The music ends, the King is seated, so are we all. "Do you mind if I smoke?" asks the woman on my left and I realize at once that I am coming down with a sore throat. "No, of course not," I smile gallantly, and she chain smokes her way through the rest of the evening.

There is a toast to the King from the Chairman of the Board of the Nobel Foundation. If my Swedish serves, it goes, in full: "To the King!" His Majesty responds with a toast to the memory of Alfred

Nobel: "Alfred Nobel!" he says. At this signal the wild rumpus starts. In spite of a prodigious array of glassware and cutlery per place setting, the meal consists of just 3 superb courses. Each begins with a musical procession of perhaps 150 waiters following the same lofty course taken by the occupants of Table A, but fanning out at the end to cover the entire floor. The exercise is carried out with military precision, under the direction of a wait-conductor positioned high on the stairway. The effect is uncanny. All across the vast hall the clatter of serving spoon against platter breaks out simultaneously and ends just as abruptly. The delivery of the first course achieves its effect simply by the shock value of seeing so many platters balanced so high descending so massive a staircase from so close to the distant ceiling. The drama of the second course is intensified by flourishes with flags and a more elaborate musical setting.

But the descent of the third course (*glace Nobel*) left me gaping in astonishment. It began, quietly enough, with a dimming of the lights. Then two great canopies flutter down out of nowhere and manage to form a kind of roof within the roof over the great staircase which is suddenly engulfed in a cascade of waist-deep white cloud that comes pouring down from the top, heralding the appearance there of two gentlemen in Turkish garb leading two proud and enormous Afghan hounds whose stately descent is followed by the costumed singers and their entourage who give us two exquisite solos, a spine-tingling duet, and a dazzling quartet. Saint-Saëns, Bizet, Gounod, and Offenbach are heard and at the moment where it has become so unbelievably beautiful to both eye and ear that one can hardly bear it, the two sopranos having ascended on a bench to a point halfway between floor and ceiling, still courageously belting it out from their perilous perch, there appear … you say, elephants?—and indeed, if the ceremony in the Concert Hall was *Die Fledermaus*, then the banquet is closer to *Aida*—but no, there appear at the top of the still smoking stairs 150 waiters bearing massive glowing trays of ice cream for the revelers below. Waiters descend, singers ascend, spoons clink, plates clatter, and a woman's voice whispers in my ear, "You're the last one—would you like everything left on my tray?" "Yes," I gasp gratefully, and soothe my aching throat with a triple portion of *panache de sorbet aux mures sauvages des champs et de parfait à la vanille* as my companion lights up yet another.

The brief after-dinner "two minute" speeches by the eldest laureate in each category are inevitably an anticlimax. Only Szymborska honors the time limit. I meditate on the narrow line dividing magnificence from bad taste. What had just taken place was unquestionably magnificent, but whether it could be pulled off outside of a monarchy is open to doubt.

The King rises as do we all, and the occupants of Table A make their exit over the reverse of the route they followed in. As the last of them mounts the great staircase, we occupants of Tables 1–65 follow them up into heaven, another enormous room covered with golden mosaics, filled with dancers and containing a cash (!) bar at the far end. The laureates are up there, still hard at work, being interviewed for TV in the back room. The rest of us sway with the crowd. Exhaustion overcomes me and we tear ourselves away for the 11 o'clock bus back to the Grand. Resisting the temptation to move on to the "Nobel Nightcap," a post-party party put on by students, we stagger home. Doffing my finery, I am Cinderella again, and lie awake the entire night, overcome by the splendor of it all and a bad cough.

Wednesday, December 11th. Stagger off to the breakfast buffet to pour bowls of hot oatmeal and honey down my miserable throat. The laureates are there, bright and chipper. They all went to the Nobel Nightcap—how could they not, poor souls. The faint of heart among them left at 3, the bravest stayed until 5. It dawns on us that if poor Professor Vickrey had not died within days of hearing of his Nobel Prize, the award ceremony would surely have finished him anyway. What will be will be. Revived by the oatmeal, I find it impossible to resist a little pate, cheese, herring, and a croissant with my tea.

I finish breakfast just in time to catch the bus for the CNN symposium in the Old Parliament building. I have checked with my map and determined that this bus ride covers a distance of no more than 300 meters, but since I don't know what door we're supposed to enter I take the bus anyway, as do all the other guests. It turns out the driver doesn't know what door either, but eventually we get in, deposit our coats, ascend another not quite so monumental staircase to find at the top a massive luncheon buffet. "Feed a cold and starve a fever" comes to mind so I load my plate very selectively. A touch of this, a morsel of that. This and that are present in enough varieties to load the plate. The herring is like butter. I ignore the blandishments of the wine attendant and wash it all down with mineral water. "Terminal dyspepsia" mutters my luncheon companion.

The CNN show seems an exercise in incoherence. A manic leering moderator who can't stay on the subject for more than two sentences, six perky—God knows how they do it—laureates in make-up, trying to say sensible things in spite of the centrifugal orchestration and lunatic premise of the whole show (the great minds of the 20th century address the great minds of the 21st). I flee the Old Parliament the moment it ends for the tranquility of the Café Access, and send a brief summary of the past 24 hours, having found from CNN the proper voice, to wit: "from the Café Access, Stockholm, [pause] I'm David Mermin." Starting to hallucinate. Passing through the Grand we encounter the astounding, indestructible laureates, smiling and spiffy in white tie and tails again. Off to the Palace they are, for their private dinner with the King. We stagger off to our own dinner. Back in the room we watch the CNN show on TV. On the screen it looks polished and witty. Amazing. They have conditioned us to accept rubbish as profound when we see it on the tube. Up all night coughing and sneezing. I have acquired the full-blown Nobel cold.

Thursday, December 12th. Photographs of the banquet arrive at the Grand after breakfast in about 25 numbered notebooks. You put your name opposite the ones you want and give the Nobel Foundation your address. They will send you a collection of photos and a bill. I realize I can go nowhere this day. Back to bed. I sleep (yes, sleep!) from noon to two. I awake refreshed. Long walk, museum, long walk back, dinner, bed. Awake all night, of course. Shouldn't take naps.

Friday, December 13th. Santa Lucia day. As we descend to breakfast we pass an ascending procession of candle-carrying maidens in white gowns. The lead maiden wears a five-candle crown (centered square). The accompanying representative of the Nobel Foundation explains as she passes us that early-morning visits in the bedroom are mandatory for the laureates (who were given Thursday night off, presumably to rest up for Friday morning) and optional for their guests, if ordered the night before. (From the concierge? Room service? Housekeeping?) Nobody told us, but never mind. They will appear at breakfast in 15 minutes. I dash down to load my plate at the buffet, dash back to the room for my camera, and as I sit down at the table it begins. The lights dim. Distant voices intoning the famous Neapolitan ditty at half tempo increase in volume until the five-candled maiden appears and leads her parade straight to our table where it stops. A 15-minute serenade follows of *a capella* renaissance tunes and Christmas ditties in candle light. The final number is a reprise of the entering *Santa Lucia*, in the course of which the five-candled maiden slowly turns and leads her crew out. As the final "Santa Lucia" ("c" pronounced like "s") fades to a whisper, the lights go back on. I wolf down my morning repast: more oatmeal with honey (for the sore throat), herring (because it's there), prosciutto (the same), assorted fruits and pastries. We finish in time to catch the bus for Uppsala.

The temperature is well below freezing for the first time and it starts to snow. We note many cars in the ditch on the way to Uppsala including two taxis, presumably on their way to the airport with passengers. Tomorrow it will be our turn to make the attempt. I waive the option of a tour of frigid Uppsala for the warmth of the auditorium in which the physics laureates give mini versions of their Nobel lectures. At the post-lecture reception, in yet another grand setting, Richardson recommends strong drink for my laryngitis. It cured his. Beer and wine flow at the sumptuous luncheon in the great hall of the Castle, but the laryngitis gets worse. Not strong enough. Doze on the bus back to Stockholm. It gets colder and colder.

Back at the Grand, the handful of remaining guests, now only enough to fill a third of a bus, set out in full formal regalia for the Lucia Dinner given by students at the Stockholm University Union. I can do nothing but croak, not entirely inappropriately as it turns out, since the pervading theme of the evening is The Order of the Ever Smiling and Jumping Green Frog. Confusion reigns at the banquet hall, a scaled-down version of Tuesday's extravaganza, but still large enough to require books and maps, which nobody thought to provide to the guests. Eventually somebody produces a list and we find our seats.

The meal is punctuated by an endless series of toasts, each accompanied by a song, lustily belted out by the toastmasters, diners, and those guests who are able to follow in the kindly provided Swedish librettos. Throat aching and mindful of Richardson's advice, I take a double dose at each of the three Schnapps toasts. The assuredly beneficial effects on the larynx are countered by an increased desire to join in the singing, which is now accompanied by the linking of arms, and a rhythmic longitudinal swaying. Wine and beer ease the pain again, ice cream appears in a great burst of sparks, the lights go out and … the soft, slow distant sounds of *Santa Lucia* float down the stairs, followed by another five-candled maiden and her white-gowned retinue. The now familiar litany of sweet and slightly mournful tunes unexpectedly modulates into a round of finger-snapping, white-gowned, hip-wiggling hot cha-cha, which just as suddenly flips back into the final worshipful round of *Santa Lucia* as they slowly and softly float away back up the stairs,

More toasts follow. The swaying is now transverse as well as longitudinal. If Tuesday's banquet was *Aida*, then this one is *Faust*. The toastmaster announces that the bus that was to meet us at midnight will not appear until 1. As it turns out this is a matter of some importance, since between midnight and 1 the culminating event of the week takes place: the induction of the laureates into The Order of the Ever Smiling and Jumping Green Frog. The bedlam of the scene in which this takes place is impossible to convey and details of the lunatic ceremony are lost in the fog, but somehow it manages to culminate in all six of the 1996 Nobel Laureates in Physics and Chemistry lined up together and uttering cries of "Ribbit, ribbit" while squatting on their haunches and jumping up and down. A fitting end to a can-you-top-this week.

Saturday, December 14th. We go out into the freezing night at 1 and there is no bus. The toastmaster has lied. There are no taxis. We try to thumb rides as the six black limos pull out, but they are all full. We peer off into the distance. Nothing. A representative of the Nobel Foundation appears, waves her magic cellular phone, and an unscheduled municipal bus pulls up at the Student Union and takes us home, free of charge. I get to bed at 3. Sleep like a log! Dream I am King. I get up at 7, sore throat gone! We take a van to the airport. I croak to the SAS representative that I do not wish to be seated two rows from the smoking section. Sorry, she says, plane full, but wait a minute. We get bumped up to business class and ride home amidst parting salvos of salmon, herring, and champagne. At home I weigh myself. Three pounds less than when I left! The final bit of Nobel magic.

36. Postscript

1. This is the original version of my Nobel diary (Chapter 19), before *Physics Today* asked me to compress it into a single *Reference Frame* column. It's more readable and gives a fuller picture of the week.

2. I'm surprised at how new ATM machines were only twenty years ago, how occupants of even the Grand Hotel had to hunt down "internet cafés" to get online, and how the only way to get news from the United States was through the *Herald Tribune*. Even using a credit card abroad seems not to be taken for granted. And cell phones are likened to magic wands. Otherwise it seems like it all happened yesterday.

37

Elegance in physics

The problem with giving a lecture on elegance in physics is that much better physicists have already said a lot about it. This is well-worn ground. The great astrophysicist Subrahmanyan Chandrasekhar began a lecture on "Beauty and the Quest for Beauty in Science" [1] by remarking that "The topic to which I have been asked to address myself is a difficult one, if one is to avoid the trivial and the banal." So I cannot even comment on the risk of triviality and banality without engaging in them.

Clearly I cannot tell you about how Maxwell's equations for the electromagnetic field used to fill an entire page and require ten different letters of the alphabet for their expression, even in empty space. But then with the invention of vectors, Maxwell's equations shrank to four lines and three letters. The added insights of relativity reduced them to one letter and one line, thereby condensing the hard-won insights of Coulomb, Oersted, Faraday, and Maxwell himself to a degree of concise precision none of them could have imagined.

Nor can I wax lyrical about the extraordinary power and economy of Dirac's formulation of quantum mechanics, which makes transparently lucid the equivalence of Schrödinger's and Heisenberg's superficially different formulations of the fundamental theory, as well as simplifying many formerly complicated calculations into a few almost self-evident lines.

And certainly I cannot say anything about the great conservation laws of energy, momentum, and angular momentum, and how they can be understood as consequences of the simple symmetries of our universe—that things happen the same way regardless of when they happen, where they happen, or how they are oriented.

All of this has been said too often to repeat any of it.

What I *can* do is tell you about a few less lofty things: attitudes of physicists toward elegance in their discipline, and a few very down-to-earth examples of what I would consider elegant.

Physicists are ambivalent about elegance. When I received the invitation to give this lecture all I could think of was "Elegance is for tailors." That succinct dismissal of the theme of this series is probably the most celebrated remark any physicist has ever made about the role of elegance in physics. But instead of thinking about my assigned topic, I began by worrying about who had said it.

It is often attributed to Einstein. If you look for it on the Web—not an elegant source of information—you turn up many such attributions. These are almost certainly wrong. The remark does not show up in any collection of Einstein aphorisms that I am aware of. And I doubt, in spite of his lack of sartorial pretensions, that Einstein would have been that dismissive of tailors. To be sure, he did say that if randomness lay at the heart of physics then he would just as soon have been a plumber, but I think he was making a distinction between science and craftsmanship, and not disparaging the plumber's skills.

It seems most likely that elegance was consigned to the tailors' shop by Ludwig Boltzmann, the great 19th-century pioneer in applying the methods of physics to inferring the behavior of bulk matter from the behavior of individual atoms, and the great proponent of the view that atoms themselves were more than a convenient fiction, but as real as things more directly accessible to our senses. Many months into my investigation, I reached the historian of science Stephen Brush. He was able to provide me with a quote from an autobiography of one E. Hlawka, cited in turn in a book on Boltzmann by A. Dick and E. Kerber. The Cornell Library possesses neither book. According to Dick and Kerber, according to Hlawka, "Ludwig Boltzmann was a decided opponent of vector analysis. He always said that mathematical calculation did not have to be elegant, that elegance should be left to tailors and shoemakers." But who was Hlawka? Who, for that matter, were Dick and Kerber? My other evidence comes from my former Cornell colleague, Michael E. Fisher,[1] who years ago pounced on me when I attributed the aphorism to Einstein, insisting that it was Boltzmann's. When I approached him in connection with this lecture, Michael confessed himself unable to cite a source. "Let me know what you learn," he said.

So when I got this news from Brush, I forwarded it to Michael as the definitive word confirming that it was indeed Boltzmann. He was not impressed. I quote from his letter:

I *personally* never had *any doubts* since I heard it *directly* from the lips of *George Uhlenbeck*—a scholar and well-steeped in the Dutch van der Waals, Ehrenfest, Kramers tradition—and on *two* or *more* occasions. Einstein's name *never* even came up in this connection—I think that was *your* idea! Perhaps Einstein even quoted Boltzmann!

And I consider the repeated word of *Uhlenbeck* as *distinctly* "more definitive" (and *direct* from "the source") than "an autobiography of E. Hlawka" (who is *he* anyway?) *reproduced* in a book *on* Boltzmann by two other authors I have never heard of!

Thanks anyway.

Quite recently I found a contribution to the Boltzmann–Einstein controversy that I dare not characterize as definitive, in a book review by Dieter Flamm, Boltzmann's grandson, who said that his grandfather "believed that matters of elegance ought to be left to the tailor and the cobbler," adding parenthetically that "Albert Einstein

---

1 More about whom in Chapter 39.

said that he adhered scrupulously to the principle in 1920 when preparing a popular lecture on special and general relativity" [2].

Whoever said it, I like the quote, because it reveals the other side of the physicists' famous fondness for aesthetic criteria of scientific merit: get the job done, by fair means or foul. Boltzmann himself had one of the most elegant insights in the history of physics. The entropy can be defined in terms of the temperature and flow of heat in a process that brings a system of fixed volume into its current configuration. He discovered that the entropy could also be expressed in terms of a purely mechanical measure of the statistical likelihood W of the system being in that configuration. The concise formula, $S = k \log W$, is engraved on his tombstone. It ranks right up there with $E = mc^2$. It is elegant in at least two respects: (1) It unites two fundamental but hitherto completely unrelated forms of description: the thermodynamic description of bulk matter, and the mechanical description of a collection of atoms; (2) Its symbolic expression is admirably brief. So there, on the tombstone of the man who consigned elegance to tailors, is one of the most elegant formulations in the history of science.

I would like to present some specific examples of what I would consider elegant, in the hope that the elegance may speak for itself, revealing its character by example.

My first two exhibits are "physicists' proofs" of the Pythagorean theorem: the square of the length of the hypotenuse of a right triangle is equal to the sums of the squares of the lengths of the other two sides: $c^2 = a^2 + b^2$. The Pythagorean theorem seems a good place to start because I was told that you would not all be physicists or engineers. Everybody has heard of the Pythagorean theorem.

I learned the first elegant proof from a book by the Russian physicist, A. B. Migdal, on *Qualitative Methods in Quantum Theory*, which confirms my sense that it belongs in a lecture on elegance in physics.

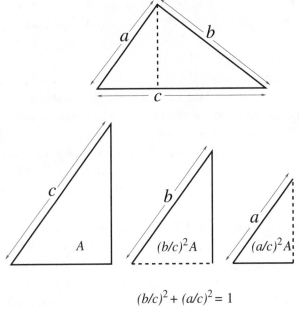

$$(b/c)^2 + (a/c)^2 = 1$$

Figure 1. Migdal's proof.

Migdal's proof starts, familiarly, with a right triangle, standing on its horizontal hypotenuse. Cut the triangle into two smaller triangles along a line through the 90°vertex perpendicular to the hypotenuse. Each piece of the original triangle shares with it one right angle and one of the acute angles. Since the sum of the angles of any triangle is 180°, all three angles of each smaller triangle are the same as the three angles of the original triangle. So each of the two pieces has the same *shape* as the original triangle: they are both scaled-down versions of the original one. So far this is boringly familiar. Many proofs start from this fact. It's like a standard opening in chess. One then labels a few line segments, sets up a few proportions, does a little simple algebra, and out pops $c^2 = a^2 + b^2$. Migdal's elegant twist is this:

Notice that if you scale down a triangle, shrinking all three sides by the same fraction f, then the area is shrunk by a fraction f^2: area scales as the square of the linear dimensions. This, already, is very much the way physicists love to think about things.

In the case at hand, the sides of the two smaller triangles are fractions a/c and b/c of the sides of the original triangle, those fractions being the ratios of the corresponding hypotenuses. So the areas of the two smaller triangles are fractions $(a/c)^2$ and $(b/c)^2$ of the area A of the original one. But since the original triangle was cut into the two smaller ones, the areas of the smaller pieces add up to the area of the original triangle: $A = (a/c)^2 A + (b/c)^2 A$, so indeed $c^2 = a^2 + b^2$.

Why is this elegant? (1) The result is important. It's hard for the inconsequential to be elegant. (2) You don't have to know much: the crucial fact that the two triangles are scaled versions of the original one depends only on the fact that the three angles of a triangle add up to 180 degrees. (3) It's unusual: you don't ordinarily build a plane geometry proof from the fact, dear as it is to physicists, that area scales as the square of linear dimensions. (4) The argument makes it absolutely clear why it's the *square* of the hypotenuse that's the sum of the *squares* of the other two sides: it's because it's the *area* of the big triangle that is the sum of the *areas* of the two little ones. So we have (1) nontrivial; (2) uncomplicated; (3) unexpected; (4) illuminating.

On the other hand the argument has some technical content. And you have to write an equation to make the conclusion explicit.

Here's an entirely different proof that is less technical than any I know. It is a solid-state physicist's proof of the Pythagorean Theorem. Solid-state physicists think a lot about the different possible ways of filling space up with identical copies of a single structure because this is the form favored by most forms of crystalline matter on the atomic scale. Two-dimensional versions of such space-filling patterns are called tilings, and can be found on many bathroom floors.

Figure 2. Tiling proof.

The solid-state physicists' proof of the Pythagorean theorem is contained in a picture of a tiled floor. Notice that there are three different sizes of squares: a big one with slanting sides, and a small and middle-sized one, with horizontal and vertical sides. There are two equally valid ways of interpreting this picture:

(a) You can think of the picture as showing two kinds of tiles: a small square and a middle-sized square, both with horizontal and vertical sides. The middle-sized square is decorated by two lines, that intersect at its center.

(b) You can think of the picture as showing only a single tile, the big square, tilted a bit. This big square is decorated by a picture of the small square, with its sides extended to form spokes that radiate out to the boundary of the big square. The corners of the big square are at the centers of adjacent middle-sized squares.

The area of one big tile is the sum of the areas of one small tile and one middle tile, since you can cut up a big tile into five pieces, one of which is the small tile and the other four of which can be reassembled into the middle tile, as Fig. 2 makes evident. Here is another way to see this. It is slightly more subtle, but much beloved by the solid-state physicist:

If you need a million big tiles to cover an enormous floor, then you can also cover it[2] with a million of each of the small and middle tiles. But if a million big tiles can cover essentially the same area as a million middle tiles and a million small tiles, then the area of the big tile must be the area of the middle tile plus the area of the small tile.

What does the fact that the area of the big square is the sum of the areas of the two smaller squares have to do with the Pythagorean theorem? Take a look at a side of a big square tilting steeply upward. Its vertical extent is from the center of a middle-sized square to the center of the middle-sized square sitting on top of the first one, so this side of the big square extends vertically by just the side of a middle-sized square. And it extends horizontally by just the amount by which the upper middle-sized square has been horizontally displaced from the lower one—i.e. by just the side of the small square.

In other words a side of the big square is the hypotenuse of a right triangle with horizontal and vertical sides, whose lengths are the lengths of the sides of the middle-sized and small squares. QED.

What makes this elegant?

(1) It's new. At least it's new for me. I made it up specially for this lecture. At the moment of discovery I felt a distinct *eureka* sensation. Given that people have been inventing proofs of this theorem for thousands of years, it can't be new to humanity,[3] but

---

2 There will be slight differences at the edges of the floor, which become less and less important as the floor becomes bigger and bigger.

3 In editing my lecture for this book I checked on Google, and indeed, it's not new.

if it's new to you that's almost as good for establishing a sense of elegance.

(2) Another criterion for elegance is that it should be the product of hard work. And indeed, it took me a couple of days to refine the argument down to the form of the picture in Fig. 2.

(3) It's pretty to look at.[4] Tilings with any nontrivial structure have a beauty to them associated with the fact that there are different ways of identifying the basic repeating unit—they have some of the charm of an optical illusion. When two different ways of picking the basic tile demonstrate a famous theorem, that lends a meaning to the beauty, which enhances it.

(4) It requires virtually no thought to grasp. You don't even have to know the famous, but, let's face it, technical fact that the sum of the angles of a triangle is 180 degrees. You don't have to know that as you rescale a figure its area scales as the square of its linear dimensions. To see the relation between the three areas you just have to look at the pieces of a jigsaw puzzle and think of sliding them around.

(5) It's surprising. You could stare at a floor tiling like that for years without noticing that it contained a proof of the Pythagorean theorem, because the relevant right triangle isn't even in the original picture. But once it's pointed out to you, you can't help seeing it after that.

Have I cheated by taking two examples of elegance in physics from geometry? I don't think so. Physical reasoning is never far from geometrical reasoning. Newton's original proof that planets move in elliptical orbits about the sun was entirely geometrical. Einstein's general theory of relativity reduced gravity to geometry (or geometry to gravitation, depending on which you think of as more fundamental). Crystallography is one of the finest

---

4 It's used to decorate the cover of this book.

flowerings of solid geometry. Much of quantum physics can be viewed as the geometry of certain higher-dimensional (or even infinite-dimensional) spaces. So elegance in geometry is very closely akin to elegance in physics.

I approach my next example with considerable trepidation. A lecture on elegance in physics should include a piece of genuine physical behavior. The elegance of the material world should be allowed to speak for itself without any intervening verbal irrelevancy. But elegance, while simple, is not easy. And the particular demonstration I want to show you is very hard to do, and not the least bit elegant if it doesn't work.

So I can only try. What I want to show you is very simple. It is marketed as a children's toy, called the Levitron™. The toy consists of two magnets. One is a small, flat, circular magnet of the kind used to stick messages to refrigerator doors. But it has a shaft perpendicular to the circle, going through its center, so it can function as a top. The other is a very large, very strong magnet, with one of its faces horizontal, with a polarity that is designed to repel the top if it is placed above the magnet.

If you just put the top above the magnet, all that happens is that the top flips over and crashes down on it, pulled there by both gravity and magnetism. But if the top is spinning, then in the classic manner of tops, it will not tip over, and will remain oriented so that it is magnetically repelled by the base. This repulsion is strong enough to overcome the downward attraction of gravity.

You might think that the top could just escape off to the side, but if it spins at just the right rate, at just the right height, in just the right position, and has just the right weight, and the base is in just the right horizontal orientation, then the magnetic force acts to push it back toward its original position no matter what direction it tries to fly off in, and the top floats elegantly above the magnet until the slight friction with the air has slowed it down to the point where it is capable of flipping over.

But everything has to be just right. I adjusted the orientation of the base and the weight of the top before you all came into the auditorium. But slight vibrations of the table top can disorient the base again and, worse, slight variations of the temperature in the room alter the strength of the magnet in the base, requiring a different set of little weights to be placed on the top. Finally, to make it even harder, it is extremely difficult to spin the top on the removable plastic slab above the magnet—it took me three days to learn how to do it, and it's not a skill I can always rely on. It's harder than riding a bicycle. This is why the Levitron™ has not been the kind of best-selling toy it would otherwise deserve to be.

But when it works, it is elegant. Or, perhaps more accurately, whether or not you find it to be elegant is a definitive test of whether or not you are a physicist at heart.

It is elegant because it embodies the basic hallmarks of elegance: (1) It is surprising: you do not often see a heavy object floating unsuspended in the void; (2) It is simple: it consists of nothing but two magnets, one of them spinning; (3) It is difficult in at least two respects: (a) It is extremely hard to get to work, and therefore deeply rewarding when it does work; (b) The explanation of how it works—why the top cannot escape sideways if its weight and rate of spin is in the right range, neither too large nor too small—is a remarkable piece of mathematical physics, going well beyond the technical capabilities of the inventors of the Levitron™.[5]

Having come up with several examples of what strikes me as elegant, I went through them to see what they have in common. I found five criteria:

5 The explanation of how it works was published a few years ago in the Proceedings of the Royal Society by the great mathematical physicist Michael Berry. He tells me that when he first submitted the paper it was immediately rejected with a brief note to the effect that "The Proceedings of the Royal Society does not publish investigations of toys." Berry is not a man to take that quietly. The Proceedings of the Royal Society *did* publish Berry's investigation of a toy, and I commend it to the engineers in the audience [3]. [To learn what happened when I attempted to do the demonstration, see note 2 in the Postscript to this chapter.]

Simple.
Not easy.
Surprising.
Illuminating.
Important.

The first two are only apparently contradictory. "Not easy" means that the simplicity must be of a nontrivial sort. To reveal the simplicity must have required an expenditure of effort. The simplicity, though you can berate yourself for not having noted it once it is brought to your attention, cannot be of an obvious kind.

The third and fourth are also complementary. For a new piece of physics to be elegant it must, to some extent, have been unexpected. But novelty is not enough. Elegance requires some degree of violating preconceptions or conventional procedures, offering a deeper insight into past ideas and practices. And it's important to be important. Something inconsequential cannot be elegant.

References

1. S. Chandrasekhar, *Truth and Beauty*, Chicago (1987). Reprint of 1979 lecture at Fermilab.
2. D. Flamm, *Physics World*, April 1999: Review of *Ludwig Boltzmann: The Man Who Trusted Atoms* by Carlo Cercignani.
3. M. Berry, *Proc. Roy. Soc. Lond. A* **452**, 1207–1220 (1996).

37. Postscript

1. This contains those parts of my lecture on elegance that were not extracted to make up Chapter 22. The talk was given in 1999 at the Weisman Art Museum in Minneapolis, as part of a year-long series on elegance in mathematics, art history, architecture, literature, fashion, advertising, and science.

2. Because of my extended warning to the audience that I couldn't guarantee that I would be able to carry out the demonstration with the magnetic top, there was an eerie silence in the room as I placed the top on the plastic slab above the magnet and got ready to try. Unexpectedly, I got it spinning on my very first attempt. I then lifted the slab, and the top jumped a little above it, just as it was supposed to do. I withdrew the slab, and there it was, floating above the lower magnet, just as advertised. I had guessed the right weights! To my surprised the audience burst into prolonged applause. Glancing at my watch, I realized that the demonstration had taken longer than I had expected and I was out of time. So I took a bow, and ended the lecture then and there.

3. Why Boltzmann's "Elegance is for tailors" is widely misattributed to Einstein is explained in the postscript to Chapter 22.

38

Questions for 2105

I've been asked to talk about questions I would ask physicists 200 years after Einstein's *annus mirabilis*, if I could be magically transported to the year 2105. Why me? Because five years ago the *New York Times* had an article about ten questions physicists would like to ask their colleagues in the year 2100. Most of those questions made me uncomfortable. They seemed temporally provincial—too absorbed with issues of the current decade or two. Consider, for example, question 4:

4. Is nature supersymmetric and, if so, how is supersymmetry broken?

and question 7:

7. What are the fundamental degrees of freedom of M-theory and does the theory describe nature?

It would surprise and disappoint me if the context and terms of such questions made sense to anybody but historians of science, after the passage of 100 years of research. The only question of the ten that struck me as reasonable was

1. Are the dimensionless parameters that characterize the physical universe calculable in principle, or are some merely determined by historical or quantum-mechanical accident, and uncalculable?

Dimensionless parameters are likely to survive all but the most radical upheavals in our conception of the world. Indeed, the concept has been with us at least since the hydrodynamicists of the 19th century, and though *fundamental* constants are a 20th-century invention, they have been with us now for the better part of a century. This question, alone among the ten, satisfies all the criteria I'll describe later.

I felt uncomfortable not only with most of the questions, but with the whole exercise. Too many unimaginable things can happen in a century to render our current concerns irrelevant or obsolete. Just think about how physics changed in the century behind us.

I write an occasional column in *Physics Today*, commenting on various peculiarities of our profession, and the pitfalls of devising questions for colleagues 100 years from now struck me as a good topic. So in February 2001 I published an essay[1] inspired by the article in the *New York Times*, giving my own list of ten questions, designed to highlight the futility of the whole exercise.

To set the stage for making inquiries a century into the future, one cannot do better than to recall the tale of Enoch Soames, as told by Max Beerbohm, in 1897. Soames is an aspiring poet. He is the author of "Negations" and "Fungoids." Both volumes have been entirely ignored by his contemporaries along with all the rest of his writings.

On June 3rd, 1897 Beerbohm runs into Soames in a Soho restaurant and they have lunch together. Soames remarks to Beerbohm that he would sell his soul to the Devil for the chance to spend the afternoon 100 years later, in the reading room of the British Museum, to learn from the card catalog whether posterity has recognized his genius.

"Permit me to introduce myself," says a "tall, flashy, rather Mephistophelian man" at a nearby table. In Beerbohm's presence, and in spite of his fervent pleas not to do it, Soames makes a pact with the Devil and vanishes from the restaurant, transported to the reading room of the British Museum on the afternoon of June 3rd, 1997.

1 Chapter 24.

Beerbohm goes back to the restaurant that evening to find out what Soames learned. Soames returns. He is furious with Beerbohm. He had failed to find a single one of his works in the card catalog. He even failed to find himself mentioned in any scholarly works on late 19th-century English literature. Indeed, he had failed to find any evidence that he ever existed.

But in the ultimate humiliation, he found himself mentioned by a late 20th-century critic, T. K. Nupton, as a fictitious character in a story by Max Beerbohm about a man who sells his soul to the Devil for a chance to spend the afternoon 100 years later in the reading room of the British Museum.

Beerbohm is so clever in exploiting the paradoxes of time travel that you have to read the story a couple of times to get all the jokes. For example, when Soames returns from 1997 Beerbohm asks what the people in the reading room looked like.

"They all looked very like one another."

Beerbohm inquires further: "All dressed in sanitary woolen?"

"Yes, I think so. Grayish-yellowish stuff."

"A sort of uniform?"

Soames nods.

"With a number on it perhaps—a number on a large disk of metal strapped round the left arm? D. K. F. 78,910—that sort of thing?"

It was even so.

"And all of them, men and women alike, looking very well cared for? Very utopian, and smelling rather strongly of carbolic, and all of them quite hairless?"

I was right every time, Beerbohm tells us.

Later on in the story Beerbohm deduces that although nobody in 1997 had ever read Soames, they had all read Beerbohm. Although he never says so explicitly, they all looked exactly as he guessed because they were, in fact, carefully reenacting Beerbohm's prediction of what the reading room was like on the precise day, June 3rd, 1997, that was specified in the story. Imagine their astonishment (described to Beerbohm by Soames) when in the midst of their literary party, Soames actually appears.

I believe that there was, in fact, a small commemorative event at the British Museum on June 3rd, 1997, but I doubt the participants put on grayish-yellow sanitary woolens and sprayed themselves with carbolic. Needless to say, Soames did not show up. Now that we have passed the precise date of Soames's appearance in the reading room, we can use his case to examine some of the pitfalls of looking 100 years into the future.

"A hundred years hence!" Soames murmurs to Beerbohm while working himself up to selling his soul to the Devil.

"We shall not be here," Beerbohm replies, briskly but fatuously.

"We shall not be here, no," repeats Soames, "but the museum will still be just where it is. And the reading-room just where it is. And people will be able to go and read there."

Well, maybe. By a strange coincidence, 1997—the exact year of Soames's visit—was the year in which the unthinkable happened. The great reading room in the British Museum, where Marx wrote *Das Kapital*, was closed, and the space was converted into an enormous circular courtyard, where you could eat lunch, and buy postcards and souvenirs. The entire library was moved to a new building near St. Pancras station, on Euston Road. The old reading room stopped functioning in October, so it may still have been limping along when Soames visited it on June 3rd.

But the card catalog is another story.

No one can condemn Max Beerbohm for failing to anticipate the electronic catalog in 1897. The conversion, in fact, is still going on and may not have been very far along in 1997. But, it was literally unthinkable in 1897, and it provides a cautionary example of the utterly unexpected things that can happen in the course of 100 years.

A complementary hazard is to anticipate changes far more radical than what actually happens. Beerbohm expected spelling reform. What Soames actually copied from the essay of T. K. Nupton was this:

"Fr egzarmpl, a riter ov th time, naimed Max Beerbohm, hoo woz stil alive in th twentith senchri, rote a stauri in wich e pautraid an immajnari karrakter kauld "Enoch Soames" - a thurd-rait poit hoo beleevz imself a grate jeneus an maix a bargin with th Devvl in auder ter no wot posterriti thinx ov im! It iz a sumwot labud sattire, but not without vallu az showing hou seriusli the yung men ov th aiteen-ninetiz took themselvz."

With these lessons in mind, I would say that a reasonable question has to meet four criteria:

(1) First of all, for purposes of this Symposium, the question should be about physics, or at least it should emerge from the attitudes and perspectives with which physicists of today view the world. Here, for example, are some questions that are not allowed, although I would like very much to ask them:

> Do people travel moderate distances in vehicles that can hold no more than four or five of them? If so, what is the energy source? If not, how, other than walking, do people make short (5–10 km) trips?

> Do you know what opera is? If so, is it still performed by singers and musicians in front of audiences?

> Are there universities at which students meet together in one place to be instructed by professors? If so, is there an academic field called philosophy?

> Did the United States of America recover from the presidency of George W. Bush? If not, did the rest of the world?

(2) The answer should, of course, be absolutely fascinating, and not just to the questioner. Think of the ways in which the young Einstein would be amazed by what we know today.

(3) The question should make sense to scientists in 2105. The danger, of course, is that a pressing question of today is likely to make sense

only to historians of science, or, worse yet, to specialists in early
21st-century issues, say, prior to the discovery in 2019 of the *chronosynclastic infundibulum*.

(4) The question should have a reasonable chance of not eliciting embarrassed giggles at the early 21st-century naivety of the questioner.

So here, after these warnings, are my own physics questions for
colleagues in the year 2105:

1. *What are the names of the major sciences? What are the names
    of the major branches of physics, if physics is still an identifiable
    science? Please characterize their scope in simple early
    21st-century terms, if you can, or try to give me a sense of why
    my ignorance makes this impossible.*

It's hard to guess what the scientific landscape will look
like in 100 years, but I can't imagine that it will look familiar. Already, for example—at least in the northeastern United
States—chemistry is trying to become a branch of biology.
At least half a dozen Chemistry Departments, including both
Harvard and my own university, Cornell, have changed their
names to Chemistry and Chemical Biology.[2] Physics, on its part,
seems bent on absorbing biophysics, and, more recently, even
economics. A Google on "econophysics" yields a modest 40,000
hits; not much compared with "quantum gravity", which yields
600,000, and supersymmetry, which gets 500,000, but the field is
new and the century is still young. "D-brane" gets only 32,000.
Looking backward rather than ahead, what would a physicist

---

2 I have been interested for several years in conceptual—perhaps even
philosophical—questions raised by the quantum theory. So when I learned that
this change of name had been made at Cornell by the chemists—sorry, I mean by
the chemists and chemical biologists—without any consultation with the rest of the
university, I asked the Dean whether it would be OK, if I could persuade my colleagues
in the Physics Department, for us to change our name into the "Department of Physics
and Metaphysics." I never received an answer.

from 1905 have made of the term "information science"? How about "nuclear engineering"?

You might complain that the real content of this question is "Tell me everything of interest." But all I'd like to learn is what unfamiliar names are going to be there. And what familiar names are going to be missing. I am certain that if physics is still a recognized field, it will have subfields whose names will have no more meaning to us than, for example, Chronosynclastic Infundibulography.

2. *Please show me a widely used, inexpensive device used that will astonish me in as many different ways as a laptop computer would have astonished a patent officer in Bern in 1905. At least some of the purposes served by this device should be as comprehensible to me as most of the uses of a laptop computer would have been to the young Einstein.*

The number of different ways in which a laptop would have amazed Einstein in 1905, is itself amazing. Forget about its primary functions. What about the material its case is made of, its cost as a fraction of mean annual income, the source of its power, the precision with which it can imitate a symphony orchestra, and its ability to show today's newspapers from all over the world and all the catalogs of all the great world libraries? Nobody imagined such a thing in 1905. Nobody today can imagine the extraordinary objects that will be found in the households (assuming there still are households) or pockets (assuming there still are pockets) of 2105.

You might object that this is a question about technology and not physics, but the technology in a laptop rests on a bedrock of fundamental physics most of which was undreamed of in 1905. I can't imagine what could rival a laptop 100 years from now in the clarity of its purpose and its ability to astonish.

Indeed, it makes me wonder if it may be a mistake to expect advances in physics-based technology in the 21st century comparable to those of the 20th century, which, after all, flowed out of discoveries in physics in the first quarter of that century that are, in some sense, unique in the history of science both for their revolutionary character and the depth and range of their techno-logical implications.

Nor does it take a prophet, or even a lot of chemists, to predict that the major advances in the 21st century will be primarily bio-logical in character. Indeed, it's from that area that I would expect the most amazing gadgets of 2105 to draw their inspiration, so perhaps this question doesn't belong on my list of physics ques-tions. On the other hand Physics Departments will probably all have become Departments of Physics and Physical Biology, in which case the question will be entirely in order.

3. *Are fundamental theories still based on superpositions of states that evolve linearly? Or have the basic principles of quantum mechanics been replaced? If quantum mechanics has survived, do people agree on solutions to the interpretive puzzles that bothered many early 21st-century physicists, or have they ceased to view them as problems needing solution? If quantum mechanics has been replaced, has the new theory clarified these puzzles, or do people find it just as, or even more, mysterious?*

Serious people are still worried by quantum mechanics. I quote one mid-century figure:

"We have always had a great deal of difficulty understanding the world view that quantum mechanics represents. At least I do, because I'm an old enough man that I haven't got to the point that this stuff is obvious to me. Okay, I still get nervous with it … You know how it always is: every new idea, it takes a generation or two until it becomes obvious that there's no real problem. I cannot define the real problem, therefore I suspect that there is no real problem, but I'm not sure there's no real problem."

I worry that this question might elicit polite bewilderment, just as a pressing ether-theoretic puzzle at the turn of the 19th century might seem not only irrelevant, but downright incomprehensible to a physicist of today.

There are two possible grounds for 22nd-century bewilderment at the question. One is that quantum mechanics will have been discovered, as Einstein always hoped, to be a phenomenology based on a more fundamental view of the world, which is more intuitively accessible. This strikes me as unlikely, particularly since John Bell showed that any such theory would have to allow instantaneous action at a distance. So while the discovery of a more fundamental view of the world during the 21st century seems possible, I'd be very surprised if the new view turned out to be more intuitively accessible then our current understanding.

An appropriate time scale for the survival of quantum mechanics is set by the fact that its basic conceptual machinery has suffered no alterations whatever, beyond a little tidying up, for 80 years. Not a bad run, when you compare what happened to fundamental knowledge between 1850 and 1930, though not close to the more than two centuries that classical mechanics remained the fundamental theory. So the persistence of exactly the same formalism for another hundred years seems at least plausible.

More likely grounds for early 22nd-century bewilderment at this question are that the theory will have survived intact, and after several more generations of physicists, chemists, and biologists (as we now call them) have worked with it, it has finally, in Feynman's words, become obvious to everybody that there's no real problem. Those early 21st-century people who believed there ought to be a better way to understand the theory will then have been consigned to the same dustbin of history as the early 20th-century ether theorists.

I hope that's not how it works out. It is, for example, now possible to articulate the nature of the wrong thinking that made relativity seem shockingly counterintuitive to many people during its early years. People had simply deluded themselves into believing that there was something called "time" that clocks recorded, rather than recognizing that "time" was a remarkably convenient abstraction—I would even say an ingenious abstraction, except that nobody set out deliberately to invent it—that enables us to talk efficiently and even-handedly about the correlation among many different clocks of many different kinds.

No comparable key to dissolving the puzzlement engendered by quantum mechanics is yet at hand. I would hope that in the next 100 years a key might be found that almost everybody would agree clarifies the character of the theory, in contrast to today's state of affairs, where no school of thought commands more than 10 percent of the population.

So although it's highly likely that if the quantum theory is still with us in essentially the same form in 2105, people will no longer be bothered by it, will that be because they've found a better way of thinking about it that removes the puzzlement? Or just that they've had a whole additional century to get used to it?

4. *Tell me about a state of ordinary bulk matter, unimagined in the year 2005, that's as remarkable as superconductivity was still considered to be in the year 2005. The extraordinary behavior should be recognizable as amazing to an early 21st-century physicist.*

2011 will be the 100th anniversary of the discovery of superconductivity by Heike Kamerlingh Onnes, and today it is at least as wonderful as it appeared to be in 1911. We now understand the mechanism of at least its most common forms, though that explanation took almost half a century to be found. More interestingly, finding the explanation took a whole third of a century after the discovery of quantum mechanics, and not for want of trying.[3]

---

3 A minor question for 2105: will people have learned that the last name of the discoverer is not Onnes, but Kamerlingh Onnes, just as the last name of the inventor of the quark is not Mann, but Gell-Mann? Because the Dutch feel no need for hyphens in compound last names, Kamerlingh Onnes seems doomed to go down as Onnes in *Scientific American* articles, textbooks, popular books by distinguished authors, and histories of science. If posterity thought of me as Min (first name Mer), while, like poor Enoch Soames, I'd be glad to be remembered at all, I'd still be pretty peeved.

Many remarkable phenomena have been predicted before they were discovered. Most famously, perhaps, was the non-Newtonian gravitational deflection of light, predicted by Einstein and preliminarily confirmed by Eddington, though the resulting hoopla was incompatible with the size of the error bars. Or, to take another example associated with Zurich, the existence of the neutrino, postulated by Pauli in 1930 but not confirmed by Reines and Cowan for another 25 years.

But I maintain that nobody could ever have predicted superconductivity. The explanation for the phenomenon—a broken gauge symmetry—is so unintuitive that it would never have occurred to anybody to think about the possibility, much less work out its observable consequences, if people had not been driven in desperation to the explanation by their efforts to make sense of the actual phenomenon. Furthermore, the mean field approximation made to extract anything useful is uncontrollable and, in many other contexts, notoriously unreliable.

In a similar way nobody could possibly have invented quantum mechanics itself, had they not been driven to it by many unambiguous but unintelligible facts. General relativity might be the unique exception to this rule, though Einstein did get an enormous clue from the Eötvös experiments and motivation from his need to extend special relativity to include gravitation.

Like superconductivity, I believe nobody could have anticipated the much more recently discovered fractionally quantized Hall effect. Without the great analog computer, Nature, to motivate our speculations and calculations, nobody would believe a word of the currently accepted explanations.

I'd love to know what other unpredictable phenomena terrestrial bulk matter is capable of. Such behavior would be an unanticipated consequence of the basic laws of quantum mechanics and electrodynamics that govern ordinary matter, but the way in which those laws gave rise to that behavior would simply be too subtle to extract, without our first having learned from Nature what it was we were looking for, and without Nature being available for us to test whatever crude or crazy ideas we came up with in an attempt to account for the phenomena.

Although only a handful of such examples were found in the 20th century, it would be strange if others didn't thrust themselves upon us as we got better and better at going to lower and lower temperatures, creating stronger and stronger magnetic fields, and fabricating devices with structures on tinier and tinier length scales.

It's an essential part of my question that the strangeness of the behavior should be intelligible to an early 21st-century physicist. This might be hard to satisfy. I'm not sure that the extraordinary character of a superconductor would have been evident to a physicist of 1811. Oersted didn't discover the action of electric currents on magnets until 1820. Ohm's law was not enunciated until 1827. Faraday discovered electromagnetic induction in 1831, all within a century of the subsequent discovery of superconductivity. The mystery of superconductivity, even if appreciated in 1811, would have been masked by a constellation of other mysteries. If the extraordinary character of the state of matter exhibited to me in the early 22nd century is that it has absolutely no coupling whatever to the *chronosynclastic infundibulum*, I'm not going to be impressed.

5. *Do time and space still play the fundamental roles they did in early 21st-century physics, or have they been replaced by more coherent, less obscure concepts?*

I am always perplexed at how people can talk about spacetime turning into a foam at the Planck scale. As already noted, the great lesson of special relativity is that the concept of time is just an extremely convenient and compact device for characterizing the correlations between the devices we are able to use as clocks, or the much broader class of physical systems whose behavior we try to correlate with those clocks. But clocks tend to be macroscopic.

They have to be macroscopic to communicate with us, which is their only purpose. Indeed, to assert that time, in quantum mechanics, refers to anything more than the time at which a state is assigned or a measurement is made, is to get into deep and murky waters.

Same problem with space. Einstein taught us that distances are best viewed as the interval between space-like separated events, and to measure these we need, for example, light signals and clocks. So when we talk about time and space at unthinkably tiny length scales, we literally don't know what we're talking about. There seems to me considerable danger here of imposing on an utterly alien realm a useful bookkeeping device we've invented for our own macroscopic convenience. The only justification for doing it (and it's not a bad one!) is that we don't know what else to do.

I confidently predict that time and space will still be with us in 2105, whatever happens to the British Library. But I wonder if they'll be in evidence at the foundations of the scientific description of Nature.

6. *Has progress been made in understanding the nature of consciousness experience or how the mind (as opposed to the brain) affects the body? Does quantum mechanics or its successor play a crucial role in that understanding? Does that understanding clarify our confusion over the meaning of quantum mechanics?*

There are those who say there is no problem of consciousness because the question doesn't make any sense. There are those who say there is no problem because the answer is obvious. Physicists further divide into those who say quantum mechanics clearly does or clearly does not have anything to do with it.

The problem of consciousness, of course, has been around for centuries. But the growing sense, at least among physicists, that science has something to say about it doesn't seem to me transparently absurd, even though no two scientists can currently agree on what that something might be.

If you're not bothered by consciousness, it's unlikely that I'll be able to explain to you why it bothers me, but let me try. The notion of *now*—the present moment—is immediately evident to consciousness as a special moment of time, or a brief interval, of order perhaps a few tenths of a second. It seems highly plausible to me that your now overlaps with my now or, if you are very far away from me, with a region space-like separated from my now. On the other hand, I can conceive of it not working this way. That your now is two weeks behind or fifteen minutes ahead of my now.

Physics has nothing to do with such notions. It knows nothing of now and deals only with correlations between one time and another. The point on my world-line corresponding to now, obvious as it is to me, cannot be identified in any terms known to today's physics. Consciousness has a particularity that seems absent from the physical description of the world, which deals only with relations. Consciousness can go beyond time differences and position itself absolutely along the world-line of the being that possesses it.

According to Rudolf Carnap, Einstein himself was bothered by "Now". Carnap reports a conversation with him in the early 1950s, in which "Einstein said that the problem of the Now worried him seriously. He explained that the experience of the Now means something special for man, something essentially different from the past and the future, but that this important difference does not and cannot occur within physics. That this experience cannot be grasped by science seemed to him a matter of painful but inevitable resignation."[4]

An even simpler example of an elementary constituent of consciousness which physics is silent on is the quality of the sensation of blueness. Physics can speak of a certain class of spectral densities of the radiation field, it can speak of the stimulation of certain receptors within the eye, it can speak of nerve impulses from the eye to the visual cortex, but it is absolutely silent about what is

---

4 I have more to say about the Now in Chapter 32.

completely obvious to me (and I assume to you)—the characteristic and absolutely unmistakable blue quality of the experience of
blueness itself.

This point—a banality among philosophers, who speak of
qualia—is extremely hard, if not impossible, to put across to many
physicists. I have sometimes managed to do it by citing a theory
I had as a child to account for the fact that different people have
different favorite colors. My idea—a kind of chromo-aesthetic
absolutism—was that there was, in fact, only one most pleasur-
able color sensation, common to all human beings, but the reason
your favorite color was blue while *mine* was red was that the sensation you experienced looking at blue objects was identical to the
sensation I experienced looking at red ones. (This same example
(complete to the choice of colors—only "you" and "me" are inter-
changed) can be found on a list of possibly meaningless questions
in P. W. Bridgman.[5])

Many people, some of them quite distinguished, have suggested
that the problem of consciousness may be related to the problem
of understanding quantum mechanics. I have little patience for
people who think that quantum mechanics may contain a solution to the problem of consciousness. Consciousness is too mysterious to find its explanation in something that simple. But I can
believe that a resolution might proceed in the other direction: if
it were possible to understand consciousness this might resolve
some of the puzzles of quantum mechanics. This has to do with
the fact that the only statements quantum mechanics makes about
the world are relational. If I view myself as a system describable by
quantum mechanics, then my state becomes entangled with anything in the physical world I interact with. My conscious perceptions, on the other hand, have a particularity that goes beyond the
correlation between those perceptions and what they are perceiving. To account for this by saying that I'm actually having all of

5 *The Logic of Modern Physics*, Macmillan (1927), p. 30.

the perceptions in a collection of parallel universes strikes me as ludicrously naive. The particularity of human experience is simply outside the scope of contemporary science, just like the nowness of now and the blueness of blue.

Many of my colleagues seem seriously to believe that when computers get fast enough and acquire large enough memories, they too will be conscious. I find this preposterous. My guess would be that in building consciousness, natural selection managed to tap into something that we don't have the slightest clue about. Perhaps the *chronosynclastic infundibulum*. The only thing I'm sure of is that we won't tap into it by building bigger and better computers. The claim that computers some day can be conscious strikes me as every bit as ridiculous as the claim behaviorist psychologists used to make that dogs cannot be.

There's a wonderful quote from Schrödinger that captures some of this feeling. Democritus, he remarks, realized that

… the naked intellectual construction which in his world-picture had supplanted the actual world of light and color, sound and fragrance, sweetness, bitterness and beauty, was actually based on nothing but the sense perceptions themselves which had ostensibly vanished from it. … He introduces the intellect in a contest with the senses. The former says "Ostensibly there is color, ostensibly sweetness, ostensibly bitterness, actually only atoms and the void"; to which the senses retort: "Poor intellect, do you hope to defeat us while from us you borrow your evidence? Your victory is your defeat." You simply cannot put it more briefly and clearly.

The risk of this question eliciting giggles—then *and* now—is substantial, but I'll take my chances. I'd love to know whether the question will be viewed as vexing, as silly, or as solved by 2105.

7. *Is the structure of matter still being probed at shorter and shorter length scales? If so, is the study still based on tracking the debris emerging from high-speed collisions? If so, how are the*

*high energies produced? If not, can you explain the alternative
investigative tools? What length scales have you reached? Has
new structure been found at all intermediate length scales?*

For nearly a century almost everything we know about matter at
small length scales has come from hurling things at each other at
higher and higher energies, so this technique is certain at least to
be remembered a century from now, even if it is no longer used.
But as the energy goes up so does the cost, and the size, both geo-
graphical and human, of the investigation.

We seem to be reaching the limits of feasibility of this method
in the early 21st century. By the early 22nd century will the
quest to probe deeper have ended, as journeys to the Moon have
ended today? Or will entirely new methods of investigation have
emerged? In the latter case, will things settle down to patterns that
persist over many length scales, as they do between 10^{-10} and 10^{-15}
meters, or will new structure and therefore new questions continue to emerge as new length scales are reached?

And will any of this—if there is any of this—make contact
with the world-view emerging from string theory? Or will the
latter be viewed as a spectacularly beautiful manifestation of
early 21st-century decadence, a kind of scientific analogue of the
Pre-Raphaelite English painters of the late 19th century?

8. *Is controlled nuclear fusion an important part of your
   technology? Are room-temperature superconductors?*

This question might well appear temporally provincial from the
perspective of 2100. But the quest for controlled fusion has been
going on for about 60 years; and the need for a clean, readily
available source of energy is more acute than ever and will only
get worse.

Materials that superconduct at liquid nitrogen temperatures and
higher have only been around for 20 years, making this part of the
question even more risky. But the five-fold increase in temperature

above absolute zero took everybody by surprise, so nobody can reasonably claim that there couldn't be another three-fold increase. The consequences for power transmission, electronic devices, and wonderful toys would be immense. Since the broader subject of superconductivity has been with us now for almost a century, and has produced one surprise after another, it doesn't seem unreasonable to guess that it could be of central importance in the technology of the early 22nd century.

So I'm fairly confident that the question will make sense in 2105. The real worry is that other technological miracles will have been devised that have made these two potential technologies so obsolete as to elicit the dreaded giggles. For example positive chronosynclastic infundibulatory feedback.

When I was a child, the most expensive department stores had networks of pipes. When you purchased something, information about the sale was written on a piece of paper. The paper, together with your money, was put into a cylinder with felt bushings at either end. The cylinder was put into a pipe and propelled by air pressure to a central office. In a little while the cylinder came back, popping out of another pipe into a basket, and there was your change. The whole thing made wonderful whooshing and banging noises—particularly at places where the pipes took 90° turns. I can imagine somebody in 1945 speculating about the amazing advances in pneumatic tubing that were likely to take place by the year 2045.

9. *Are there quantum computers that can factor thousand-bit*
   integers? What else are they used for? Do most homes have one?

This is by far the most rash of my questions. The whole subject of quantum computation is so new that it all may well have evaporated by 2015. The question would then make sense in 2105 only to a few historians of science.

There are two major obstacles to the existence of quantum computers in 2105. The first is technological. To factor a thousand-bit integer on an ideal quantum computer, you need at an absolute minimum two thousand Qbits. Quantum error correction, which it's hard to imagine won't be an essential component of such a device, multiplies this number by a factor of seven, so we're up to a couple of tens of thousands of two-state systems, whose interactions with external fields and whose pairwise interactions with each other must be controlled with exquisite precision, and whose interactions with anything else must be reduced to an extremely low level, so error correction can be effective. At the moment it's a technological triumph to produce half a dozen Qbits—almost enough for one error-corrected logical Qbit.

Is this a worry? Half a century ago I had a summer job at IBM on Madison Avenue in midtown Manhattan. My job was to write a program to invert a matrix of complex numbers on the very newest computational wonder, the IBM 704. We wrote in assembly language. We were told that somebody was working hard to develop a more intuitive but incredibly inefficient language called FORTRAN, which was being written for people who were too stupid to write their programs in the vastly more efficient assembly language.

The 704 occupied half a city block. If somebody had told me that within half a century there would be vastly more powerful computers that you could carry in your pocket that cost less than ten 1956 dollars and ran off a battery the size of an American penny that lasted for many years, I would have sent them off for psychiatric care. So I am inclined to dismiss my first concern, as symptomatic of the chronic lack of vision that has afflicted me all my life.

The second obstacle may be more serious. The only practical task that a quantum computer would be exponentially better at than a classical computer (according to our current understanding of classical algorithms) is efficiently determining the period of certain periodic functions, notably $f(x) = a^x(\text{mod } N)$. Other applications are closely related to this one. Lov Grover's famous search algorithm is completely different, but it gives only a square-root speed-up over a classical search.

The ability to solve this problem efficiently permits one efficiently to factor N, and this, in turn, compromises the security of the widely used RSA scheme for encrypting secret messages. (RSA encryption can also be broken directly by an efficient period-finding machine, without the detour into factoring.)

While only a fool would expect that by 2105 the moral level of humanity would have risen to a stage where it was no longer necessary to keep secrets, it isn't rash to expect that some time before then, other forms of encryption will be found that are not vulnerable to an attack by period-finding. Indeed, individual Qbits, carried by the polarization states of photons, provide a method for replenishing one-time code pads that's already proved feasible with contemporary technology, and it's entirely plausible that in another 100 years this will be the dominant means for exchanging secret messages, in the (unlikely) event that no better classical procedure has been discovered.

Should one be discovered, the financial backing for research into quantum computers will suffer a precipitous decline, making their practical realization even less feasible. This would be a pity, since quantum computation is one of the most beautiful, surprising, and illuminating applications of quantum mechanics to have arisen in the second half of the 20th century.

10. *Have intelligent signals of extraterrestrial origin been detected?*

There's little to say about this one. It's not terribly expensive. The technology for searching is bound to get better and better. So I hope somebody keeps on looking for at least another century, not so much because I think the odds of success are high, but because it would be so wonderful if we did. It would be a great comfort to know that if we managed to eliminate intelligent life from Earth—either by deliberate acts of war or terrorism, or our inability to control the mindless greed that threatens to make the planet uninhabitable—that we would not have destroyed something unique in the universe.

While 10 is the canonical number of questions to ask, with the example of Enoch Soames in mind, I cannot refrain from concluding with an eleventh:

11. *Is my book on special relativity,* It's About Time, *published by Princeton University Press in late 2005, still in print? Or, better yet, will it be in the process of being reissued in a special centenary edition?*

Of course the question rashly assumes there will continue to be such things as books. I fear it is hopelessly naive. More likely, there will be 10-petabyte crystals people carry in their pockets containing permanent copies of everything ever published, updated continuously, and readable through special eyeglasses powered by ambient illumination. Yes, my book will have survived. *Everything will have survived.*

But this view of the future, like all the views I have expressed here, is probably hopelessly wrong. So I would urge you to watch your bookstores in late 2005 and buy a copy while you still have a chance.

38. Postscript

1. Unlike Chapter 37, which is the rest of a lecture that subsequently shrank into Chapter 22, Chapter 38 is a lecture into which Chapter 24 eventually expanded. The occasion for the talk was a conference, "Physics in the 21st Century – 100 Years after Einstein's *Annus Mirabilis*," held in Zurich in 2005. The organizers asked if I could give a version of my year-2000 column, updated five years to "What I'd Like to Know about 2105."

2. Since my original column had been severely criticized by David Gross, the impresario of the conference of string theorists whose silly questions had inspired my original column, I welcomed the opportunity to make my point more fully. What I didn't anticipate was that Gross would be in the audience in Zurich. Two minutes into my talk he rose to denounce me all over again. As it became apparent that he had a lot to say about what was wrong with my column, I quietly sat down in the front row, waiting for him to finish. This put the audience solidly on my side, and when I was able to resume, they enjoyed every minute of the remaining fifty-eight. I recall Jennifer Chayes (who, it can now be revealed, was my hostess in Los Angeles who knew how to spell "Lagrangian" as told in Chapter 1) laughing so uncontrollably at my remarks about Bush the Second, that I had to ask her permission to continue.

3. I'm able to report that in 2015, ten years after this lecture, my 2005 relativity book is still in print.[6] And people are still interested in quantum computers. And no serious ones have yet been built.

6 Indeed, a German translation became available in 2015.

PART FIVE

Some People I've Known

39

My life with Fisher

Many years ago I was writing a talk, "My Life with Landau," for a conference[1] commemorating the 80th anniversary of the birth of the great L. D. Landau [1]. I knew I was going to have to deliver it before an audience that included Michael Fisher, and I found, to my distress, that as I sat there at the keyboard the image of Michael kept intruding on my thoughts, questioning my assumptions, denouncing mean field theories, and otherwise disrupting my concentration, in the way that we have all come to know and love. Finally, to chase him away, I wrote "Some day I would like to give a talk on 'My Life with Fisher'" and strangely enough, that got rid of him. But I've known ever since that the time would come when I would have to pay for that liberating moment.

I first heard of Michael Fisher in 1963, at the beginning of my postdoctoral year at La Jolla.[2] I met another young postdoc, Bob Griffiths, and in response to the intellectual sniffing out that goes on on such occasions, Griffiths let it be known that what he was up to was proving that the free energy of a spin system exists. "That it *what?*" I said. "That it *exists*," said Griffiths firmly. "I'm using some ideas I got from Michael Fisher." Well, I thought, this Griffiths seems like a nice guy anyway. And I decided that this mentor of his, this Fisher, must be a man with deep philosophical interests—a sort of Plato of thermodynamics.

---

1 Footnote: The same conference plays a similar role in Chapter 42.
2 See Chapter 40 for more about that year.

I didn't hear of Fisher again until I got to Cornell in 1964 and Ben Widom told me one day that Michael Fisher was coming for a visit. "That's nice," I said, and remembering him as Griffiths' mentor, looked forward to meeting such a quiet and contemplative man. The visit lasted over 20 years, and turned into the most wonderful thing that has happened to me in my professional life.

Let me trace for you Michael's trajectory through the acknowledgments sections of my publications. He first shows up at the end of the 1966 paper in which Herbert Wagner and I give our version of Hohenberg's theorem. Herbert and I had tried to explain to Michael that an argument of Pierre's could be adapted to prove that there could be no spontaneous magnetization in the two-dimensional Heisenberg model. I hadn't known Michael for very long at that point, and one of the first things I learned was that you should think twice before claiming to *prove* something in front of a man who encourages postdocs to show that the free energy *exists*. He didn't believe a word of it. Spectral functions, indeed! How did we know those frequency integrals even converged? It soon became evident that we were dealing with a man who knew nothing about quantum field theory, didn't care one bit that he didn't, and was convinced that we would be better off ourselves to forget it. Immediately.

So in the face of this astonishing attack, we worked backwards, unbundling our result from the conceptual wrappings in which it was enshrouded by some of the great thinkers of the previous decade, peeling off layer after layer, day after day, in the face of unrelenting skepticism, until finally we had it down to a trivial statement about finite dimensional matrices.

And then an astonishing change took place. "Publish!" he practically shouted, "it's very important!" and having learned what it was like to be at the end of a Michael Fisher attack, I suddenly learned what it was like to have him on your side. Freeman Dyson came to town. Michael introduced us. "Mermin and Wagner have proved that there's no spontaneous magnetization in the two-dimensional Heisenberg model," Michael proudly informed him, as Herbert and I basked in his admiration. "Of course there isn't," Dyson responded. "But they have *proved* that there isn't," Michael insisted. One Dyson eyebrow may have moved up half a millimeter in response. No matter. I was hooked on arguing with Michael Fisher. My life would never be the same.

Here are some later acknowledgments:

In a 1967 footnote: "The analysis given here was constructed at the suggestion and with the vigorous assistance of M. E. Fisher." It's a footnote rather than an acknowledgment, because in those days they wouldn't let you say anything human in an acknowledgment.

In 1968 we read: "M. E. Fisher's insistence on the difficulty of specifying a criterion for crystalline ordering led me to discard several earlier versions of the argument."

In 1976 we read: "We are indebted to M. E. Fisher for lending us what seems to be the only copy of de Gennes' book now in Ithaca."

In 1977: "The importance of these considerations was brought home to me by a ferocious lunchtime discussion with M. E. Fisher."

In 1979: "It was M. E. Fisher who first suggested and repeatedly insisted that I should publish my lecture notes, but I am not sure he deserves thanks for this."

Finally in our 1976 book Neil Ashcroft and I, after thanking 47 alphabetically arranged colleagues, devoted a whole paragraph to No. 48:

One person, however, has influenced almost every chapter. Michael E. Fisher, Horace White Professor of Chemistry, Physics, *and* Mathematics, friend and neighbor, gadfly and troubadour, began to read the manuscript six years ago and has followed ever since, hard upon our tracks, through chapter, and, on occasion, through revision and re-revision, pouncing on obscurities, condemning dishonesties, decrying omissions, labeling axes, correcting misspellings, redrawing figures, and often making our lives very much more difficult by his unrelenting insistence that we could be more literate, accurate, intelligible, and thorough. We hope he will be pleased at how many of his illegible red marginalia have found their way into our text, and expect to be hearing from him about those that have not.

I call your attention to our characterization of Michael as a gadfly. It was only after coming to know Michael that I fully understood what the Athenians meant when they called Socrates a gadfly, and shortly after that I also began to understand why they had made him drink the hemlock. I think most readers understood what we meant by "gadfly," until the book started being translated into other languages. It was Michael himself who reported to me, with only the slightest tinge of acidity, that a Japanese friend had nervously asked him why our preface called him a "small, but loud and annoying insect."

The Russian translator simply gave up and replaced "gadfly" with "pedant." I knew the Polish translator had taken a more serious approach to the problem, but I never got around to figuring out just what Michael was called in the Polish translation, until, in preparing this 70th birthday speech, I sought help from Wojciech Zurek:

Dear Wojciech,

Could you help me with a translation? In our book Neil Ashcroft and I refer to Michael Fisher as "gadfly and troubadour." In the Polish edition "gadfly and troubadour" comes out as *ciety jak osa i wesoly jak trubadur*. My theory is that "gadfly" has become *ciety jak osa* and troubadour has been expanded to *wesoly jak trubadur*. Am I right and can you give me a translation of these phrases? I have to give an after-dinner speech at a banquet in Fisher's honor.

Here are some excerpts from Zurek's reply:

The translation is not bad, though it does change the meaning of the original phrase a bit: *ciety jak osa* means "ready to bite like a wasp." You could also say *giez* (which is literal for "gadfly"), but you would not say this about anyone in an after-dinner speech in his honor....

On the other hand, *wesoly jak trubadur* (literally "gay as a troubadour") probably changes the intent. I am guessing *wesoly* was added for reasons of symmetry, to balance the *ciety*.

All the best,

Wojciech

P.S. Why are you giving your after-dinner speech in Polish?

In 2001, in reassuring defiance of all the reckless gossip about our book getting out of date, the first German translation appeared. Here Michael is our *Freund und Nachbar, Troubadour und lästiger Zeitgenosse*, so in certain German circles, Michael is now becoming known as a troublesome contemporary.

The translators of the French edition in 2002, like the Russians several decades earlier, also gave up, replacing "gadfly and troubadour" with *rigoriste et poète*. But the Portuguese translation of 2011 has *moscardo e trovador* which, as far as I can tell, gets it exactly right!

I was away from Ithaca on sabbatical in Rome during the great 1970–71 revolution in our understanding of phase transitions. Michael told me all about it when I got home. What particularly impressed me was this: In the years before that *annus mirabilis* Ken Wilson[3] would drop by my office every year or two and say mysterious things about phase transitions. When we were both 17 we

3 See Chapter 41.

had the same German teacher as freshmen at Harvard, so I knew he was pretty smart, but I really thought he was losing his marbles with this talk about rolling balls uphill with just enough energy so they almost made it all the way to the top. And then all this sloppy stuff in momentum space. He didn't even know how to write proper integral signs. So I was really amazed to come back home and find that Michael—a man who was interested in whether the free energy *existed*, mind you—had just waded right in, and was able to explain to me what Ken had been unsuccessfully trying to tell me. He had even learned about Feynman diagrams.

But in the middle of all that unrigorous slop, he never forgot about his high standards. He gave a wonderful colloquium on what mathematical physics was all about. This is a pretty hard thing to do in a colloquium, but he managed to make it absolutely gripping. I'd just come up with my own definition of the difference between mathematical physics and theoretical physics that I was planning to use in a colloquium I was to give at Princeton the following week, so I tried it out on Michael after his lecture: The distinction, I told him, was not to be found in the physics, but in the sociology of physics: theoretical physics was done by physicists who lacked the necessary skills to do real experiments; mathematical physics was done by mathematicians who lacked the necessary skills to do real mathematics. Michael was not amused. "I'd advise you not to say that at Princeton," he snarled. I did anyway. It nearly set off a riot.

He was right, but the nice thing about Michael is that he is always ready to give you advice about anything whatsoever, and if you don't take his advice, he doesn't hold it against you. He never forgets, of course, that you didn't, and is quite willing to remind you, very sympathetically, when you get into trouble because you didn't. The reason he is so good at giving advice is that he thinks very seriously about everything, and always seeks out the best advice himself. He once asked me how I would find out where to buy a typewriter in New York City. I said I really couldn't tell him, because all I would do would be to ask my father-in-law. What's his name, he asked? The next time I spoke to my father-in-law he remarked that a strange thing had happened. A man with a very loud voice had phoned him in his law offices and asked where to buy a typewriter. "What did you do?" I asked. "I told him, of course," said my father-in-law impatiently. He was like Michael in some ways.

We all know that Michael has strong opinions about everything, but what always fascinates me about Michael's opinions is that although they are the strongest and most forcibly argued opinions I have ever encountered, I can never predict in advance what direction they will point in. Closely related to this is the most profound unwillingness to settle for things the way they are that I have ever run across.

What does Michael Fisher do when he checks into a hotel room for a night? He rearranges the furniture. He'll rotate the bed 90°, put the TV in the closet to make more room on the desk, carry the desk over to the window to get more light. He is an inspiration to me. Often I find it valuable to ask myself at difficult moments, what would Michael do?

Let me give you a recent example. A few years ago I was at the annual meeting of the Danish Physical Society which took place at a small conference center south of Copenhagen. Each conferee had a little apartment with a tiny attic. Downstairs was a living room and bathroom. Up a narrow ladder was a built-in bed in a room with no light. Since one used the apartment only at night this was an irritating arrangement. It was clear to me what Michael would have done. So I dragged the mattress and bedding down the ladder, remade the bed on the living room floor, and never climbed up to the attic again. This solution would not have occurred to me if I had not asked myself "What would Michael do?"

The next day various Danish conferees complained about the arrangement. Ah, I said, under such trying circumstances you should always ask yourself what Michael Fisher would do. That night the air was filled with mattresses hurtling down ladders. I believe there is now a flourishing "What Would Michael Do?" movement among Danish physicists.

Sometimes the answer to "What would Michael do?" is clear, but one lacks the courage to do it. Here is a good example:

Michael and I were flying from Copenhagen to Ithaca together. The flight stopped in London, but after we reboarded and the door had shut, the plane was slow to leave the gate. As time went on it began to look more and more like we would miss the Ithaca flight. When the likelihood began to approach certainty, Michael, muttering that there was no reason to spend the night on a bench at Kennedy when he had a brother-in-law in London, rose from his seat and announced to the flight attendant that he was getting off. "You can't," she said. "Yes I can," said he. "We're about to depart," she said. "You've been saying that for an hour and a half," said he. "Michael, sit down," I said. "Shut up," said he. And he strode past her toward the closed door. "Open the door and let me out," he said in the general direction of the door. "Your baggage is on board," said they. "Hold it for me in New York, I'll pick it up tomorrow," said he.

And then something happened that I wouldn't have believed. The door opened, a ramp appeared, and shouting back to me (who had for some time been pretending he was a complete stranger) "See you tomorrow in Ithaca!" off he strode. Immediately thereafter the door closed, and the plane took off, landing in NY just in time for me to make the Ithaca flight which had, as usual, been delayed. I got home without any waiting at all.

I conclude the story of my life with Fisher with the tale of how Dorothy and I came to own a microwave oven. Six years ago I agreed to spend three months in Leiden as Lorentz Professor. My immediate predecessor in that position was Michael E. Fisher. I remarked to a friend that Michael would be a tough act to follow. No, he said, on the contrary: following Michael had to be the easiest way to be Lorentz Professor because, as he put it, "Nothing you ask of them will seem unreasonable."

When we were first shown the Lorentz Professor's apartment, I was surprised to see a microwave oven in the kitchen. We had never had one ourselves, so I remarked on what a well-appointed kitchen it was. "Yes," our host said, "the microwave is quite new. We just got it last year." Apparently Michael, on first being shown the apartment, had looked it over and said, "What, no microwave?!" So for three months we enjoyed the Fisher microwave. When we got home I looked around our kitchen and said "What, no micro-wave?!" We have had one ever since.

The Lorentz Professor sits at Lorentz's old desk. Attached to it is a brass plaque stating that between 1878 and 1912 the desk was used by H. A. Lorentz. At Lorentz's desk was a chair. Attached to it I found a brass plaque stating that in 1994 the chair was used by M. E. Fisher. Whatever Michael thought of H. A. Lorentz, he apparently did not admire his notion of what made for a decent desk chair. As I result, I sat very comfortably for three months at the Lorentz desk in the Fisher chair. There cannot be many who, for so long a period, have been made *more* comfortable by Michael. Gadflies do not make people more comfortable.

I have to say that life in Ithaca without that kind of excitement is a shadow of what it used to be. Michael lived just down the street from me. A lot of physicists were in the neighborhood. As you walked down the street looking at the mailboxes you would read Berkleman, Mermin, Widom, **FISHER**, Webb. On the other hand life in Maryland seems to have heated up. After Michael had moved there and bought a new house, I asked how things were going. "Not well," he said. "Why?" I asked. "We decided to move the walls out three feet," he said. "Which walls?" I asked. "All of them," he said.

I conclude this birthday speech as I began it, with another acknowledgment. This one is from my scientific contribution to the Michael Fisher 60th Birthday Festschrift ten years ago:

I would like to thank God for arranging our lives so I could spend over two decades with Michael Fisher at Cornell, and His servant, the National Science Foundation, for supporting this investigation through Grant No. PHY9022796.

I would be delighted to thank the National Science Foundation for supporting this latest tribute to Michael Fisher under Grant No. PHY0098429. But I'm not sure God's servant would consider it an appropriate use of His resources, so I won't.

Reference

1. "My Life with Landau" can be found as Chapter 4 of N. David Mermin, *Boojums All the Way Through*, Cambridge University Press (1990).

39. Postscript

1. This is based on an after-dinner talk given at the 70th Birthday Conference for Michael Fisher at Rutgers University in December 2001. It incorporates additional text from the after-dinner talk I gave at Fisher's 60th Birthday Conference at the National Academy of Sciences in Washington, D.C. in 1991. A version appeared in *Journal of Statistical Physics*: "My Life with Fisher," *J. Stat. Phys.*, **110**, Nos. 3–6, March 2003, 467–473.

2. Michael Fisher is the man who taught me, among many other things, to number all displayed equations. See Chapter 6.

40

My life with Kohn

In 1963 I applied for academic positions in the United States from Birmingham, England, where I was finishing two postdoctoral years in Rudolf Peierls' wonderful department.[1] In those early post-Sputnik halcyon days you then applied for assistant professorships and collected the offers that came in reply. Peierls had suggested that I send such a letter to Walter Kohn. Disappointingly, he offered only a two-year postdoc. But I had never been to California and Peierls clearly had a high regard for the man, so this otherwise noncompetitive proposal was tempting. I asked the people at Cornell, who did come through with a faculty position, whether I could defer it for two years to do a postdoc in La Jolla. They said two would be too many but one would be OK. Walter said one was fine with him, so in August 1963 I showed up in La Jolla.

Walter was not there. He was finishing a sabbatical in Paris. But there were several wonderful postdocs, the Physics Department was still located right on the beach, Dorothy and I found a house in Del Mar on a cliff overlooking the Pacific for $128 a month, and life was good. (I had written a letter from Birmingham to an acquaintance from college who was then a postdoc in La Jolla, asking what it was like there. All I remember from his reply was "Volleyball is standard on the beach at noon.")

Eventually the moment of truth arrived. The boss returned. The holiday threatened to end. Walter invited me to his office to say hello. It was immediately clear that this was a kind, charming, witty

1 For my life with Peierls, see Chapter 42.

man. After we had exchanged pleasantries, he told me about a little theorem he and Pierre Hohenberg had proved back in Paris. The proof was one of those clever three-line arguments that wouldn't have occurred to me if I had thought about it for a hundred years, but was utterly simple and transparent when Walter laid it out in front of me.

He asked me to think about how to generalize the theorem from the ground state to thermal equilibrium. I returned to my office to consider it, and quickly realized that a new variational principle for the free energy that I had formulated in Birmingham for an utterly unrelated purpose, seemed to be tailor-made for generalizing the Hohenberg–Kohn theorem to nonzero temperature. It took me less than an hour to check that their proof did indeed go through in exactly the same way if the ground-state variational principle they used was replaced with my thermal equilibrium variational principle.

So I went back to Walter's office and knocked on the door. Here's how you do it, I said. He seemed somewhat taken aback and before I got very far into my explanation he offered to explain again what the problem actually was that he wanted me to work on. He treated me to his beautiful and transparent argument again. I said yes, that was what I had understood the argument to be (he really had explained it very well the first time) and my point was that it worked just as well when the temperature was not zero if you used the variational principle I was trying to tell him about. He was deeply skeptical. Slowly it dawned on me that this was the problem he had hired me to spend the year working on.

It took me a day to convince him that I had indeed answered his question. Then he was very pleased and I, needless to say, was ecstatic. Throughout childhood I was the last to be picked when baseball teams were being formed. I could never get my bat to make contact with the ball. But one day, by sheer chance, I got the bat in the right place at the right time and the ball went sailing over the heads of the outfielders. I must have been ten years old. It was a magical moment. Now I was 28, at the beginning of my career, and it had happened to me again. Never in the long professional life that followed did I ever again have as glorious a moment.

Since I had finished the year's work, Walter encouraged me to think about whatever I felt like thinking about—pelicans, whales, body surfing, physics. We become good friends. As we said good-bye at the end of the year he said, "By the way, when you get to Cornell, write up that theorem. Some day it may be important."

Frankly, I wasn't so sure. But I did feel he ought to get something out of having maintained me in that semitropical paradise for a whole year, so I dutifully wrote a very short paper in Ithaca [1]. When writing this memoir in 2003 I subjected my Hohenberg–Kohn Corollary to a citation search. The first decade after it appeared (1965–74) bears out my doubts. The number of citations per year ranged from 0 to 2, most of them traceable back to Walter and his collaborators. Then, ever so slowly, Walter's "some day" started to dawn. During the next six years (1975–1980) annual citations varied between 3 and 7. Between 1981 and 1993 there were between 17 and 23 citations per year (except for a 12 in 1982). And from 1994 to 2002 (with Walter's 1998 Nobel Prize in chemistry smack in the middle) there were as many as 40 and never less than 28. It may now be my second most cited paper.[2] Not bad for an hour's work!

Honesty compels me to acknowledge that Walter has a different view of this history. He maintains that it took me 24 hours, not just one, to do his postdoctoral project. Although his ability to recall ancient events is normally phenomenal, in this case I am rather

---

2 In editing the text for this book of essays I checked again. In 2015 it is still my second most cited paper, though number three is closing in on it. In 2014, the last complete year before writing this, 50 years after my hour's work, it received 75 citations.

sure that doing the job took an hour; it took me the next 23 to convince him that I really had done it.

Walter was not only directly responsible for my finest hour, but he was also present at my finest half-hour, a quarter of a century later. In 1989 I gave a lecture in St. Louis on "quantum nonlocality" at a March APS meeting with the nonstandard title "Can You Help the Mets by Watching on TV?" To my amazement and the astonishment of several of my friends, two thousand people showed up. It was in a hotel ballroom but there was standing room only.

Among the multitudes, to my surprise and delight, was Walter, who had watched, bemused, as my interest in foundations of quantum mechanics slowly developed during the 1980s. Every time I met him during those years he'd want to know just what it was that bothered me about good old quantum mechanics. He was always very nice about it, saying he just didn't get it—what exactly was the problem? I knew that deep in his heart, though he was much too kind to come out with it directly, he didn't really think this was a fit preoccupation for one who had once been capable of doing a year's work in a day. (No, Walter, in an hour!)

After a wild half-hour talk and an even wilder question period, Walter made his way through clusters of fiercely arguing people up to the podium to say hello. He shook my hand warmly, beamed his wonderful smile at me, and said "I still don't get it."

I cannot claim to have been present at any of Walter's finest moments, but I was there for two quite fine ones. The first was his 60th birthday celebration in Santa Barbara, in 1983. Vinay Ambegaokar and I sang a long song, setting his CV to music in the form of a well-known Gilbert and Sullivan number. The banquet audience, provided with copies of the words, served as chorus. Well do I remember the voice of Pierre Hohenberg, who served as master of ceremonies, lustily belting it out.

The opening verse celebrated Walter's very first publication:

When I was a lad I thought a lot
About the heavy and symmetric top.
I thought so much they published me
In the 'Merican Mathematical Society ....

There was a verse on Kohn anomalies, the second line of which was inspired by a wonderful phrase from Walter's paper:

Amongst the phonons I could see
The image of the surface of the Fermi sea!
[Chorus:] Amongst the phonons who but he
Could dare to see the image of the Fermi sea?

After the performance Walter remarked that he had always been quite proud of that line, as indeed he should be. I have always admired him as a prose stylist as much as I admire him as a physicist—whoops, chemist!

I was also present, along with about five thousand others, at Walter's meeting in Rome with Pope John Paul II in the year 2000. To appreciate the moment from my perspective you have to know the old joke about Louie:

Louie knew everybody. "There ain't nobody I don't know," he boasted. "C'mon," said Al, "you don't know the Pope." "Want to bet?" said Louie. So off they flew to Rome, where we find them standing amidst a huge throng in Piazza San Pietro, waiting for the papal blessing. People in the crowd who walk past them are saying "Hi Louie," "Wie geht's, Louie?" "Ciao, Louie," etc. "Stay here," says Louie to Al, and disappears into the crowd. Some time later the Swiss Guard appear on the steps, there's a fanfare of trumpets, and out comes the Pope onto the balcony of St. Peters, arm in arm with Louie! An enormous cheer goes up from the crowd and a monk, standing near Al, turns to him and shouts over the roar, "Who's that guy up there with Louie?"

At the end of a week of sixty simultaneous conferences in all areas of human knowledge, in celebration of the Jubilee Year 2000, there was a final super-plenary session at the Vatican. A group of us walked there from our hotel and just as we came upon an enormous array of outdoor tables alongside St. Peter's, loaded with five thousand little plastic cups of espresso ("With the compliments of His Holiness"), we noticed that Walter had disappeared. There were rumors that he had been siphoned off for something special, but nobody knew for sure. The rest of us were conducted to seats fifteen or twenty rows back from the stage in an auditorium that made my ballroom in St. Louis look like an intimate seminar room. After a couple of hours of inspirational warm-ups, His Holiness appeared on the stage, listened to four people each give a one-minute summary of 15 conferences, and then gave some remarks of his own putting it all into perspective.

That being done, a couple of dozen cardinals formed a line in the central aisle and one by one walked up onto the stage and, kneeling before the Pope, Kissed the Ring and were blessed. They were followed by an even larger collection of Archbishops, who were followed by an endless string of Bishops. Then the laity joined the procession and suddenly we noticed the back of a familiar head slowly approaching the stage. Could it be our missing Walter? What would he do? Nervously we awaited the meeting. Yes, it was Walter! When the time arrived, standing before the Pope, he Shook the Hand and launched into what, in comparison to the preceding brief encounters, could only be called a little chat. At this precise moment Giovanni Bachelet, a native Roman, who had encouraged several of us, including Walter, to participate in this extraordinary week and who had made it clear to us that he was a devout member of the Church, called out to me across several aisles: "David, who's that guy up there with Walter?"

I conclude with the latest version of Walter's musical CV, produced for his 90th birthday celebration in 2013. The song was sung by Vinay Ambegaokar and Jim Langer. I accompanied at the piano. The bracketed lines in boldface were to be sung by the guests at the party, who were provided with copies of the text, with the accented syllables indicated, lest they tie themselves up in knots.

I'm particularly proud of the rhyme in the first two and a fraction lines of verse 7.

The Musical Curriculum Vitae of Walter Kohn, sung in celebration of his 60th Birthday, March, 1983, revised and updated for his 90th, June, 2013.

(1)
When Í was a lád I thóught a lót
Abóut the Héavy and Symmétric Tóp.
I thóught so hárd they públished mé
In the 'Mérican Math'mátical Socí-etý.
[The Américan Math'mátical Socí-etý.]
But Máth-e-mátics wéren't for mé
So I wént to get a Hárvard Physics P´h.D´.
[No máth was nót his cúp of téa.
So he sóught to get a Hárvard Physics P´h. D´.]

(2)
In Cámbridge Tówn I máde my wáy
To whére the Schwínger students láugh and pláy.
I fóund it récreátionál
To invéstigate some méthods variátionál.
[He invéstigated méthods variátionál.]
My trí-al fúnctions sóon taught mé
All the fácts about collísions of light núc-lee-ée.
[His trí-al fúnctions sóon decrée-
-d'Everythíng about collísions of light núc-lee-ée.]

(3)
The Schwínger stúdents Í had séen
Preóccupíed themselves with fúnctions Gréen,
But fúnctions Gréen were gréy to mé
So I sáid good-býe to quantum fíeld theorý.
[Yes, he sáid good-býe to quantum fíeld theorý.]
Oh nó more núc-le-óns for mé:
I prefér to think abóut cohesive énergý.
[For núc-le-óns do nót bring glée
When a lád can think a-bóut cohesive énergý!]

(4)

Then Píttsburgh néxt becáme my hóme
And sóon my íntellect begán to róam
To ée-lectrónic própertíes
Like dónor lévels near impúritíes.
[**And accéptor lévels near impúritíes.**]
Amóngst the phónons Í could sée
The ímage of the súrface of the Férmi séa!
[**Amóngst the phónons whó but hé**
Would thínk to seek the ímage of the Férmi séa?]

(5)

But whén La Jólla cálled to mé
I léft the Ínstitute of Cárnegíe.
And thére I cáme to únderstánd
That one néed not cálculate each énergy bánd.
[**No, one dóesn't have to cálculate each énergy bánd.**]
For ée-lectrónic própertíes
Are úniversal fúnctions of their dénsitíes.
[**Yes évery síngle própertý**
Is a fúnctional of ée-lectrónic dénsity!]

(6)

In-hóm-o-géneous ée-lectróns
Are at súrfaces and líe along the válence bónds.
Whatéver quéstion yóu may ásk
I've a fúnctional desígned to do that véry tásk.
[**Yes his fúnctionals worked brílliantly at évery tásk.**]
They wórked so wéll they enlísted mé
To bé the first Diréctor of the ÍTP´.[3]
[**He shówed a fíne propénsitý**
To presíde at the foundátion of the ÍTP´.]

---

3 The Institute for Theoretical Physics in Santa Barbara, California.

(7)

And só I móved to Sánta Bár-
B'ra ás the Réigning Physics Théory Czár.
The fórmer ÍTP´ todáy
Has been súp-ple-ménted with the létter K.[4]
[Yes it's béen expánded by the létter K´.]
But mý K síts upón the wáll
Of the ácademic búilding that is cálled Kohn Háll.[5]
**[Yes hé's the óne who's knówn to áll
As the ánimating spírit of his ówn Kohn Háll.]**

(8)

My fúnctionáls perfórmed so wéll
That in nínety-eight the Swedes produced a Príze Nobél.
The Nóbel Príze is véry fíne
But I dídn't want to gét it until nínety-níne.
[He postpóned his Swedish vísit into nínety-níne.]
And althóugh I have a Hárvard-Physics-P´hD´,
They awárded me my Nóbel Prize in chémistrý.[6]
**[Notwithstánding his Hárvard-Physics P´hD´,
They préferred for him to háve their Prize in chémistrý!]**

(9)

So nów I'm nínety withóut much fúss
And of phýsics *and of chémistry* eméritús.
The jóurney thére was ráther lóng:
My Currículum Vítae makes a níne-verse sóng.
[His Currículum Vítae needs a níne-verse sóng.]
From a lád with a tóp I have grówn to bé
A vénerable fígure at KÍTP´.
**[From a héavy tóp with sýmmetrý
He has rísen to the tópmost rank of chémistrý.]**

---

4 Thanks to a grant from the Kavli Foundation.
5 Since the renaming some of us like to call him "Kohn of Kohn Hall."
6 Many eminent physicists have suffered this indignity.

Reference

1. N. D. Mermin, "Thermal Properties of the Inhomogeneous Electron Gas," *Phys. Rev.* **137**, A1441–3 (1965).

40. Postscript

1. This was prepared for a volume of essays about Walter Kohn on the occasion of his 80th birthday.[7]

2. I have added to the essay the text of Walter's musical CV, which was written to celebrate his 60th birthday, and then updated for his 90th birthday celebration.

7 *Walter Kohn: Personal Stories and Anecdotes Told by Friends and Collaborators,* Matthias Schemer and Peter Weinberger (eds.), Springer (2003).

41

My life with Wilson

I met Ken Wilson in 1952, when I was 17 and he was 16. We were both freshmen in the Harvard class of 1956, which produced an exceptionally large number of well-known physicists. As I remember Ken and I were in the same introductory German class, but it might have been the required freshman composition course, which met in the same building. What I remember for sure is that I got an A and Ken got a B. This was the only time in my life that I understood something better than he did.

What I remember most about Ken from 1952 is that although he was 16, he looked about 13. I don't think he looked 16 until he got tenure at Cornell. An important part of his youthful look was that when somebody said something that really pleased him, his face would light up with such a sweet smile that it warmed your heart, like the first smiles of a baby.

Indeed, there was a performance at the 1964 Cornell Physics Department Christmas party, the year I got here, in which graduate students sang satirical songs about their Professors. What they sang about Ken was based on a song from South Pacific ("Bloody Mary is the one I love, now ain't that too damn bad!") It went "Kenny Wilson is the cutest Prof, but mesons ain't much fun." Mesons may indeed not have been much fun, but within ten years Ken was to provide theoretical physicists with the best opportunities for fun that they had since the invention of quantum mechanics.

Ken and I were math majors at Harvard. He had a formidable reputation as a mathematician and was at least two years ahead of me in the math curriculum. It being Harvard, however, by general

341

agreement, which I believe Ken subscribed to, he was only the second-best mathematician in our class. I found this abundance of superior talent discouraging. To escape from mathematicians who were too smart to compete with, I stayed at Harvard for graduate school, but switched to physics. Out of the frying pan into the fire.

In 1959 Ken reappeared in Cambridge as a Junior Fellow. Much to my surprise, he too had become a physicist, irritatingly fast, at CalTech. But, also to my surprise, he seemed especially interested in electronic computers. This was still a time in which most of those who actually did computations used slide rules or, if we were really up to date, noisy electro-mechanical calculators. I thought it was a funny direction for such a talented mathematician to be moving in. Nobody had told me about von Neumann.

Ken was the first person I ever heard utter the phrase "electronic mail." "What's that?" I asked. He explained. What a silly idea, I thought. If such a thing had ever existed in the past, it would surely have been wiped out by the invention of the telephone. I think of this as my email moment. It was not the only email moment I had with Ken.

When I arrived at Cornell in 1964 as a new Assistant Professor. Ken had already been there for a year. I took his presence as a warning that I had gotten in over my head.

The two decades starting in the late 60s were a very special time to be in the Cornell Physics Department. A member of an external visiting committee evaluating the Department remarked to me after they had talked with a room full of graduate students: "They think they're in Valhalla!" Not one, not two, but three different future Nobel Prizes were being worked on.

There was the discovery of superfluidity in helium-3 by Dave Lee, Doug Osheroff, and Bob Richardson. There was an extended visit to Cornell by Tony Leggett, during which he put together, before our very eyes, his beautiful explanation of all the strange nuclear magnetic resonance signals observed in that superfluid.

And there was Ken.

Michael Fisher and Ben Widom ran a very lively weekly seminar on phase transitions. It started at lunchtime, and went on until everybody except Michael was exhausted. The only allowed medium of communication was chalk and blackboard. Interruptions of the speaker were strenuously encouraged.

In 1966 a preprint arrived from Leo Kadanoff. I volunteered to give a report on his manuscript at the Fisher–Widom seminar. I remember my title: "My Early Impressions of Kadanoff's Partial Justification of Widom's Conjecture on the Homogeneity of the Equation of State Near the Critical Point."

I didn't think much of Leo's partial justification. As I studied it, it struck me as a sloppy piece of work. He started with a well-defined problem and then proceeded to butcher it, replacing detailed local configurations by gross averages, making unbelievably crude approximations to force those averages to look like the original problem, and then repeating the whole messy business over and over again.

In those days there was a wall between particle theorists and condensed-matter theorists that was rarely crossed, but to my surprise Ken showed up for my talk. Even more to my surprise, he seemed to take Kadanoff seriously. About a week later he showed up at my office, ostensibly to talk about Kadanoff's paper. But what he actually had to say was that the crucial thing about the whole business was to think about a ball rolling up a hill with just enough energy to get exactly to the top. I humored him. "Yes Ken. Very nice Ken," I said as I ushered him out the door. It was another email moment for me.

The rest, of course, is history. I consider it one of the great experiences of my life to have had a front-row seat at the revolution, and to have had Michael Fisher and Ben Widom at Cornell to explain to me over the next few years, with great skill and patience, what Ken had been trying unsuccessfully to tell me at the beginning.

A superb former Cornell professor of mathematics, Mark Kac, visited us often during the Valhalla years. He famously said, "There are two kinds of geniuses: the ordinary and the magicians. An ordinary genius is a fellow whom you and I would be just as good as, if we were only many times better. There is no mystery as to how his mind works. Once we understand what they've done, we feel certain that we, too, could have done it. It is different with the magicians. They are … in the orthogonal complement of where we are and the working of their minds is for all intents and purposes incomprehensible. Even after we understand what they have done, the process by which they have done it is completely dark."

Mark was 20 years older than Ken and me, and although he clearly admired Ken, he was as suspicious of his theory of the critical point as I was of Kadanoff's "partial justification." Mark was as wrong as I was. Ken Wilson was a magician. It was a privilege to have known him.

41. Postscript

1. These were the opening remarks at the Ken Wilson Memorial Symposium, held at Cornell in November 2013. Leo Kadanoff was to have opened the Symposium, but he called in sick a day or two before. I was not on the program, but volunteered to replace him.

2. I remarked at the beginning that I thought I had known Ken longer than anybody else in the room, unless his brother David happened to be there. He was not. I asked if there were any cousins there. None. Then Ben Widom called out from the audience to say that he had known Ken longer than I had. It turned out they had met a year before I arrived at Cornell. Ben

didn't realize until after my talk that Ken and I had known each other for twelve years before we were reunited in Ithaca.

3. A version of these remarks appeared as "Early memories of Ken," *J. Stat. Phys.* **157**, 625–627 (2014); also in *Ken Wilson Memorial Volume: Renormalization, Lattice Gauge Theory, the Operator Product Expansion and Quantum Fields*, B. E. Baaquie, K. Huang, M. E. Peskin, and K. K. Phua (eds.), Singapore: World Scientific (2015), Chapter 4, pp. 83–85.

42

My life with Peierls

In 1988, on the 80th anniversary of his birth, there was a conference in Tel Aviv in memory of Lev Davidovich Landau. I announced at the beginning of my own talk—"My Life with Landau"—that I was going to break new ground as a scientific expositor, by following the practice of humanists and reading from a manuscript that I had prepared in advance.

To my surprise Rudolf Peierls, who was sitting in the front row, shook his head emphatically. Professor Peierls disapproves, I told the audience. "Yes," he said, "this is not a good way for a physicist to give a talk." "Well," I said, "it's better than what most physicists do now, reading their talk from plastic sheets while projecting it on a screen." He looked at me with that sweet half-smile of his and said "That is also a bad way to give a talk."

In preparing this lecture I reread *Bird of Passage*, Peierls' 1985 memoirs [1]. There I discovered that these views, like everything else I ever heard from him, were not casually put forth. They were based on experience and reflection. He says (p. 256):

It is perhaps a weakness that I find it very difficult to lecture from a written text. I cannot just read the text, because reading and sounding as if one knew what one was talking about is a difficult art, which I never mastered. Writing out the text and memorising it is equally difficult. I tried on some occasions to write out a draft, and then to follow the general trend of it, without trying to reproduce the text verbatim. With this, too, I ran into trouble, because in writing the draft I had used some turns of phrase to lead from one thought to another, and I clearly would have liked to repeat these, but as I had varied the wording in talking they would not fit in.

Because of this disability I learned to speak freely...

Nevertheless, as in Tel Aviv in 1988, so in Santa Barbara in 1997, fully mindful of the fitting-in problem, I will try to follow a written draft without trying to reproduce the text verbatim. Today it will be easier because, most regrettably, Peierls is not here with us to explain exactly what is wrong with this approach.

In my own career as a physicist and lecturer I've been inspired by two great expositors: Feynman and Peierls. I wouldn't attempt to rank them, because they followed complementary approaches to the problems of exposition: Feynman was Dionysian; Peierls, Apollonian. But behind the celebratory flamboyance of one and the serene clarity of the other, the substance was often remarkably similar. Here, for example, is a sentiment I have always associated with Feynman in the 1960s:

Sometimes people inclined to philosophic generalizations attempt to show that the basic laws, such as those of Newton, are really self-evident from some alleged general principles of which we can be sure even apart from any knowledge of the world around us. This is a rather risky undertaking because the most beautiful derivation of a law of nature from abstract principles must collapse if it turns out that the law itself has to be modified in the light of later discoveries. If it should ever come about that our knowledge of the laws of physics turns out to be complete, then it will, of course, be quite safe, for anyone who wishes, to prove that they could not possibly be any different from what they are.

This is Peierls, writing in 1955 [2]. More about his attitude toward philosophers later.

To establish that my admiration for Peierls as an expositor was not just cooked up for the topic assigned me on this occasion, let me cite the final paragraph of the Preface of my *Solid State Physics* book with Neil Ashcroft:

One of us (NDM)… is also deeply indebted to R. E. Peierls, whose lectures converted him to the view that solid-state physics is a discipline of beauty, clarity, and coherence.

My coauthor, by the way, was appalled when I produced this sentence. He clearly viewed it as saying more about the nature of a Harvard physics education in the 1950s than about Peierls. I insisted on keeping it. Our compromise was that he then added

The other (NWA), having learnt the subject from J. M. Ziman and A. B. Pippard, has never been in need of conversion.

In the late 1950s and early 1960s, when you finished your Ph.D. thesis at Harvard you so informed the National Science Foundation. They then provided you with a two-year postdoctoral fellowship which enabled you to live opulently in Europe, at the prevailing rate of exchange. You then informed Aage Bohr of your situation and went off to Copenhagen.

This system broke down in 1961. All went according to plan until Bohr replied that they had run out of office space on Blegdamswej and could not accommodate me. I was stunned. "What now?" I asked my advisor. "Birmingham." he replied instantly. Visions of *matjes* herring gave way to visions of boiled cabbage. Except for meeting my wife Dorothy, it was the most fortunate thing that ever happened to me.

I had indeed been taught solid-state physics as a melange of cook-book recipes, not worthy of being taught in a Physics Department. So it was a revelation, when I arrived at Birmingham as a postdoc in 1961, to attend Peierls' lectures on solid-state physics and discover that he offered as gripping and ingeniously constructed a detective story as any of the other subjects Harvard saw fit to teach in its Physics Department. My aim in writing my own solid-state physics text seven years later was simply to merge the gracefulness and intellectual power of

Peierls' own book with the wealth of phenomenology and pretty pictures in Charles Kittel's.

But remarkable expositions in Birmingham weren't limited to Peierls' solid-state-physics course. There was the weekly "mathematical physics" seminar, which in spite of its title was extremely wide-ranging. Regardless of the subject of the talk and regardless of his apparent state of wakefulness or somnolence during the lecture, Peierls would invariably begin the question period by managing to ask a question that was, in fact, an extremely polite but concise and lucid revelation to the audience of what it actually was that the speaker had been trying to say—or what it was the speaker ought to have been saying. Any one such performance was impressive enough, but to see it happen over and over again, week after week, was truly dazzling.

He refers to this extraordinary gift of his only once in his memoirs (p. 265). He recalls in 1953 arriving early for his own talk at a conference in Kyoto. There he found himself listening to Lars Onsager expounding what turned out to be the most powerful diagnostic tool of metals physics—the manifestation of electronic band structure in the de Haas–van Alphen oscillations of the magnetic susceptibility.

[Onsager's] mind worked so fast and was able to deal with such subtle arguments that he had difficulty explaining his ideas even to normally competent theoretical physicists.... As I listened to his talk I understood his trick immediately, because I was very familiar with the background, but I also realised that most of the audience had understood nothing. So I abandoned the talk I was going to give, and instead gave an explanation of what it was that Onsager had done.

That's scientific exposition at the heroic level.

Peierls' extraordinary talent for exposition was not confined to seminars or the classroom. We all went to the Birmingham Staff House each day to partake of their appalling lunches. He had a theory about English cooking too (p. 91):

[I]t would be undemocratic for the *cook* to impose his or her taste on the *guests*, so things are boiled until only a neutral matrix remains, to which the guest can give any flavour by adding salt, pepper, horseradish, mustard, ketchup, and so on.

We subjected ourselves to this culinary style rather than bringing sandwiches to work, not out of masochism. The inducement was that either at the meal itself or at after-lunch coffee in the lounge, one might run into Peierls and be treated to one of his endless supply of stories, about physics or physicists or, best of all, both at once. All those stories about physicists are now preserved for posterity in his memoirs. He reports Genia objecting when he embarked on the project. "It will not be Literature!" As usual, in my own experience, she was right. Here is how he describes his method:

I decided to take the memories as they came, and present the reader with a random selection from these different types of people and places.

And that's what it is. All his favorite stories, from what it was like to work with Pauli or with Bohr, to tales of his own research students, postdocs, or visitors in Birmingham or Oxford. For anyone who spent any time with Peierls the book is vividly evocative—for me of happy times in the Birmingham Staff House and wonderful holiday celebrations at Carpenter Road.

Everything is there except his technical physics stories. Unlike too many physicists, Peierls clearly viewed it as bad manners to produce these in the mixed company of physicists and nonphysicists. But although they play a minor role in his memoirs, they are preserved in his two "Surprises in Theoretical Physics" books [3], which capture wonderfully both the style of his lectures, as well as some of the tales he loved to tell when there were only physicists around the table. You can see him in these essays carefully laying out all the evidence and then putting the pieces of the puzzle together. Over and over again he pulls out from under the rug for display and disposal a variety of embarrassing questions that everybody else had been happy to sweep out of sight. The scope of his attention is breathtaking.

You can't talk about Peierls as an expositor without also mentioning his essays that graced the pages of the *New York Review of Books* in the 1970s and 1980s. They were, for me, the primary reason for maintaining my subscription to that periodical in spite of its weakness for articles that were overly precious or verbose. Once or twice a year Peierls would be there. His essays were exemplary. They got straight to the point, they wasted no words, they drew on his personal experience without any of the manifestations of self-display one has come to associate with the *New York Review*, and when he had made all his points, he stopped.

In preparing this talk on Peierls as an expositor I checked in the Cornell Library Catalog to see if there were any books of his I was unaware of. This is how I rediscovered his book for non-scientists, *The Laws of Nature*, written, as the dedication suggests, to help put Gaby and Ronnie through college:

To the Treasurer of Somerville College, Oxford, and the Bursar of Gonville and Caius College, Cambridge, but for whom this book would not have been written.

I had read and enjoyed it in Birmingham in 1962, and forgotten all about it. It was a revelation to read it again, after several decades of my own struggles to make some of the ideas of physics accessible to people with no technical training or mathematical sophistication. I was amazed to discover some of the points of view I thought I had painfully constructed on that field of battle, gracefully and lucidly tossed off by Peierls. I mention only his discussion of Newton's laws of motion. It is the only one I know at any level that disentangles what is definition from what is empirical fact, and does it simply, clearly, and quickly.

In the preface to *The Laws of Nature* I also found an explicit statement of what I had viewed as my own hard-won discoveries about the real reason why teaching is so important at a research university (p. 7):

I often found that the experience of stating an argument in simple, nontechnical terms helped me to understand it more clearly myself than when I had looked at it only in its mathematical form.

and (p. 8):

I have, in fact, in advanced lectures to students, used with advantage some of the illustrations worked out in the course of writing [for nonscientists].

Another aspect of *The Laws of Nature* that startled me today were Peierls' views on scientific knowledge. They ought to be required reading for contemporary sociologists and philosophers. I knew already that Peierls (like Feynman) took a dim view of the work of philosophers of physics. It was he who first commended to me Pauli's characterization of philosophy: "The systematic abuse of a nomenclature invented for that purpose." On one of Peierls' visits to Ithaca I remarked to him that I was on a committee to establish a new program in the History and Philosophy of Science. Ah, he said, the history of science is an interesting and important discipline. But, he added, there is no such thing as the philosophy of science.

In spite of that view (or perhaps because of it), in *The Laws of Nature* he displays a deeper and more concise insight into the philosophy of science than most of its more celebrated practitioners. Take this post-Kuhnian view of the nature of scientific revolutions, from his introduction to *The Laws of Nature*, seven years before Kuhn's *Structure of Scientific Revolutions* and a decade and a half before the birth of the sociology of scientific knowledge:

At this point the reader perhaps objects: if physicists are really as careful as that in forming their opinions, how is it that their views of the fundamental laws can be subject to such drastic changes as in the case of relativity, which destroyed the results of Newton's mechanics, or in the case of the theory of light quanta replacing the wave theory of light? How could scientists believe for centuries in theories which were later found incorrect?

This is an important question. The answer is that physics (or for that matter any other empirical science) cannot claim that its results have absolute and final truth. What can be claimed is that we have a picture which is in many respects very close to the truth, so close in fact that for most purposes the differences are unimportant. As time goes on the laws of physics which we have formulated are subject to tests of greater accuracy, or are tested in regions which are much wider than those for which the law was originally developed. In this way a need for revision is sometimes discovered, and the law has to be widened so as to include phenomena which were not previously suspected; but its first form remains a valid law for all practical purposes for the phenomena for which it was first devised.

I was particularly struck by Peierls' view of the role of Michelson–Morley in supporting the theory of relativity, since I have recently made the identical point in a series of exchanges with sociologists over the last year in *Physics Today*:[1]

From reading this chapter [on relativity] so far the reader will have got the impression that we have built an elaborate structure of reasoning on one experimental fact, namely the negative result of the Michelson experiment. However reliable the experiment, and however attractive the general principle of the independence of all laws of physics of the state of motion of the observer, one would not have accepted such far-reaching conclusions without a great deal of further support. In fact other physicists have claimed from time to time that on repeating

1 See Chapter 18.

the Michelson experiment they did find a positive answer, but in the meantime other evidence for the theory of relativity had become so strong that we would have no cause to change our views if some fundamental flaw was discovered in Michelson's reasoning.

The Cornell Library catalog also yielded up a small book by Peierls—just a pamphlet—that I was entirely unaware of. Called "Is there a crisis in science?", it's the text of the Joseph Wunsch Lecture, given in Haifa in 1971. I had known that he had been a Wunsch Lecturer because I was invited to be one myself in 1992. Before accepting, I asked Peierls about his own experience. "Do it," he said, "they take very good care of you. Genia and I refer to it as our Wunsch fulfillment." He was right.

I was delighted to find in this essay one of the few biting judgments Peierls seemed willing to set in writing:

It is not just a question of two cultures, as that great master at oversimplification, C. P. Snow, would have us believe.

He also offers an interesting view on the remuneration of scientists: (p. 6)

It has often been stressed that scientists should be looked after, but that it would be a disaster if their salaries were substantially above those of other people, because then some would turn to a scientific career as a means of making money... I do not, of course, want to imply that scientists should be seriously underpaid. It is right that scientists should bring some small sacrifice for the enjoyment which their profession can bring them....

He explicitly addresses the problems of scientific exposition:

[W]e must be ready to communicate our findings in such a form that someone not trained in science but with a serious will to understand, can follow. This is a great art, and it is becoming more difficult with the present trend to specialization and complication. Scientists who have not learned to communicate with their colleagues down the corridor who work in a slightly different field are not in a strong position when they try to communicate with the general public.

And

We should encourage more scientists to give time to the task of writing, of presenting areas of science in accessible form. There is a tradition that a good scientist should produce new results, and that writing is a less constructive activity, to be left to those who are past their active research period, or who are not quite able enough to do good research. This may have been true at some time; it is not true today.

Peierls was one of the few physicists of his generation who retained an active interest in contemporary discussions of the foundations of quantum mechanics. He had firm views, including a strong distaste for the term "Copenhagen interpretation." In *Surprises* he writes (p.26):

In recent years the debate on these ideas has reopened, and there are some who question what they call "the Copenhagen interpretation of quantum mechanics"—as if there existed more than one possible interpretation of the theory.

Whatever you call it, his brief discussions in *Surprises* and *More Surprises* constitute the most concise, coherent, and thought-provoking exposition of—I can, alas, say it safely here—the Copenhagen interpretation that I know of. I have been urging philosopher friends to have a look for many years.[2] I don't know if any of them have.

---

2 I say more about Peierls' views of quantum mechanics in Chapter 33.

Let me conclude with a story about Peierls, relevant to this question, which it seems to me is in the spirit of the stories Peierls himself loved to tell about others.[3] I used to be fond of quoting one of Aage Petersen's reminiscences about Niels Bohr:

When asked whether the algorithm of quantum mechanics could be considered as somehow mirroring an underlying quantum world, Bohr would answer, "There is no quantum world. There is only an abstract quantum physical description. It is wrong to think that the task of physics is to find out how nature is. Physics concerns what we can say about nature."

Several years ago I read this statement of Petersen at the MIT physics colloquium, and Viki Weisskopf was instantly on his feet. That was outrageous, he said; Bohr would never have said such a thing. You should not spread such slanders. This had a powerful effect on me. I stopped quoting Petersen.

Several years later in Oxford, the last time I saw Peierls, I wanted to ask him what Petersen might have had in mind and started to tell him this story. I only got as far as the pseudo-Bohr quote when Peierls' eyes lit up. Yes, he said emphatically, that's exactly the kind of thing Bohr used to say. When I told him the rest of the story he just shrugged his shoulders. So I started using Petersen's quote again. But I also concluded that not only is there more than one interpretation of quantum mechanics: there is more than one Copenhagen interpretation.

I very much wish that Rudi Peierls were still among us today, to tell me clearly and precisely, without reading from a text, exactly what is wrong with that conclusion.

---

3 Chapter 25 expands on this anecdote.

References

1. R. Peierls, *Bird of Passage*, reissued in Princeton Legacy Library, Princeton University Press (2014).
2. R. Peierls, *The Laws of Nature*, Scribners (1956), p. 275.
3. R. Peierls, *Surprises in Theoretical Physics*, Princeton University Press (1979); and *More Surprises in Theoretical Physics*, Princeton University Press (1991).

42. Postscript

1. In 1997 there was a gathering at the Institute for Theoretical Physics in Santa Barbara, part memorial, part reunion, to reminisce about Rudolf Peierls, who died in Oxford in 1995. I was asked to talk about Peierls as a writer. Since I have admired his writing throughout my professional life, it was the right assignment. In the course of putting my talk together I found many wonderful Peierls quotations that I had either forgotten or never known about.

2. Peierls, more than anybody else, taught me how to write physics.

PART SIX

Summing it Up

43

Writing Physics

One of the questions asked of lecturers in this series[1] is "What first drew you to your discipline?" What first drew me to physics was magic. It came in two varieties: relativity and quantum mechanics. I know that the magic of relativity had grabbed me before I was 16, because I remember my first day of high-school physics in 1951. The teacher was a tight-lipped gentleman who, it was rumored, had risen to the rank of Colonel in World War I. He liked to throw hard rubber erasers at people he thought were dozing.

"Physics," the Colonel told us, "is about laws that govern the behavior of matter. There is, for example, the law of the Conservation of Mass." My hand shot up. "Doesn't relativity say that mass is *not* conserved?" There was a long terrifying silence. "I don't know anything about relativity," snarled the Colonel at last. "Do you?" I never again inquired about anything not in the textbook: *Modern Physics*, by Charles E. Dull.

So I must have known something about relativity when I was sixteen. The magic of it was this: If you could move at 99.98% of the speed of light, then in a little over four years you could go four light years, and get to the nearest star. But—here was the magic part—you would be only a month older when you got there. Same thing on the way back. When you got home everybody would be eight years older, but you would have aged only two months. If you

---

1 The Knight Distinguished Lectures in Writing in the Disciplines, Cornell University, 1998–9.

did it three or four times you could come back younger than your own kids!

Just as amazing, if on the way out in your spacious mile-long rocket, you passed another one-miler on the homebound run and measured its length as it flashed past, you would find it to be only one foot long, and everybody in it, flattened to the thickness of sheets of paper. And, most mysterious of all, the occupants of the home-bound rocket would find that you and your rocket were correspondingly squashed. How could this be? How could each of two rockets be shorter than the other? I desperately wanted to understand.

Although I took a whole course in relativity in graduate school, I didn't really understand until after I arrived at Cornell in 1964 as an Assistant Professor of Physics. I was given my first semester off from teaching, the better to prove my prowess as a hot-shot researcher. (This was a big mistake since I was looking forward to teaching and, in my new environment, feeling distinctly unhot.) In November, however, I was asked to give a substitute lecture in the big introductory physics course. I attended the lecture before mine. It was on relativistic length contraction and I didn't understand a word of it. Fortunately I had the weekend to think it over, and realized that the reason I didn't understand was that most of what the professor had said was wrong. So I figured out how to say it right, and began my lecture with a delicately worded "review." From then on I was hooked on the teaching of relativity.

The following semester the Chair of my department decided that besides teaching a high-powered graduate course, I should teach a course for high-school teachers in a science-education program in Cornell's College of Agriculture. The course was supposed to be about a new set of teaching experiments. "Here," he said, handing me a key. "They're all in a filing cabinet on the third floor of Rockefeller Hall."

I climbed to the third floor and opened the cabinet. It contained what looked to me like a pile of undifferentiated junk. What to do?

Fortunately the Chair had also put me on a committee to look into how to improve the teaching of high-school physics. So I came to the committee meeting with an agenda, inspired in part by my run-in, fourteen years earlier, with the Colonel. After due consideration, our committee concluded that the most effective way to improve high-school physics was to incorporate relativity into the curriculum.

Thus empowered, I announced to my dozen high-school teachers that I would provide keys to the cabinet for anybody who wanted to play with the experiments, but that I was going to teach them relativity. They were delighted. A month into the semester I received a letter from the Department of Science Education denouncing me for dereliction of duty; I sent them back a copy of our committee report and never heard from them again.

That was in 1965. My lecture notes for the high-school teachers became a book [1], from which I sporadically taught relativity to nonscientists at Cornell until about 1990, when I decided that the book no longer worked for me. Partly this stemmed from the increasing discomfort I felt pretending I was the same person as the owner of the brash narrative voice from the pre-revolutionary side of the 1960s. More important, though, was my discovery, long after my class with the high-school teachers, that writing relativity wasn't as easy as I once had thought, as successive generations of Cornell undergraduates were bringing vividly to my attention. Now I'm in the process of writing a competing book [2].

What makes writing relativity so tricky is this: Built into ordinary language—in its use of tenses, for example—are many implicit assumptions about the nature of temporal relations that we now know to be false. Most importantly, Einstein taught us in 1905 that when you say that two events in different places happen *at the same time* you are not referring to anything inherent in the events themselves. You are merely adopting a conventional way of locating them in time, that can differ from other equally valid conventional assignments of temporal order which do not assign the same time to the two events.

This error—the implicit assumption that the simultaneity of events happening in different places has more than a conventional meaning—can infect statements that seem to have nothing to do with time. Detecting the hidden presence of time can be challenging. Suppose, for example, somebody has a garage with doors at the front and the rear on a circular drive. Whether or not "the car is shut in the garage" can be a matter of convention. For implicit in being "shut in the garage" is that neither door is open. For the car to be shut in the garage, both doors must be closed *at the same time*. If "at the same time" is a matter of convention, then under appropriate conditions it can be a matter of convention whether the car was ever shut in the garage.

Time also makes an implicit appearance in the correct assertion that the mile-long rocket is only a foot long when moving, because the length of a moving rocket is the distance between two places. The first place is where its front end happens to be at some moment; the second place is where its rear end is *at the same time* that its front end is in that first place. If "at the same time" is problematic, then so is the length of anything that moves.

Language evolved under an implicit set of assumptions about the nature of time that was beautifully and explicitly articulated by Newton: "Absolute, true, and mathematical time, of itself, and from its own nature, flows equably without relation to anything external ..." Lovely as it sounds, this is incorrect. Because the Newtonian view of time is implicit in everyday language, where it can corrupt apparently atemporal statements, to deal with relativity one must either critically re-examine ordinary language, or abandon it altogether.

Physicists traditionally take the latter course, replacing talk about space and time by a mathematical formalism that gets it right by producing a state of compact nonverbal comprehension. Good physicists figure out how to modify everyday language to bring it into correspondence with that abstract structure. The rest of them never take that important step and I would argue that, like the professor I substituted for in 1964, they never really do understand what they are talking about.

The most fascinating part of writing relativity is searching for ways to go directly to the necessary modifications of ordinary language, without passing through the intermediate nonverbal mathematical structure. This is essential if you want to have any hope of explaining relativity to nonspecialists. And my own view is that it's essential if you want to understand the subject yourself.

The other magic that first drew me to my discipline was quantum mechanics. When I was fourteen or fifteen I read a book by George Gamow [3], which I learned about from an editorial in *Astounding Science Fiction*. I've never been so excited by a book. It was my first exposure to the amazing facts of relativity. It also taught me some remarkable mathematics that I could actually see for myself made sense. This gave me some confidence that the author could be right about the other things too. Gamow also talked about quantum mechanics.

There was a car in a garage again. This time all the doors were shut all the time, so there was no possible ambiguity about "the car was in the garage." And yet, Gamow said, it was possible, without opening any doors, or damaging the car or the garage, that the car might subsequently appear outside the garage. He was using this as a metaphor for the kind of radioactivity in which the nucleus of a heavy atom can shoot out a small fragment of itself, even though there is not enough energy for the fragment to penetrate the confining walls of the nucleus that contains it.

Quantum mechanics is much stranger than relativity. The strangeness arises again out of the incompatibility of ordinary language with the actual facts. But unlike relativity, nobody has clearly identified the traps in ordinary language that make it so difficult to talk correctly about quantum mechanics, and nobody has come even close to finding a way to use ordinary language that eliminates the perplexities.[2] As with relativity, physicists have discovered a mathematical formalism that gets everything right, leading to a state of nonverbal comprehension. Nobody has yet figured out how do that in ordinary language, so trying to write about quantum mechanics without using the mathematical formalism is an exquisite challenge.

Suppose, for example, you have some stuff, and you put it into some kind of testing device that always responds, by signaling Yes or No. Suppose when you test some particular stuff the device says Yes. You might be tempted to conclude that this was the kind of stuff that produces the answer Yes when tested, but that, of course, is more than you are entitled to say, since the response of the device might have little or nothing to do with the character of the stuff you just tested.

But surely you are entitled to conclude that this was not a specimen of a kind of stuff that *must* test No. After all if the stuff did require the answer to be No, the answer couldn't have been Yes. But it was Yes. So it was a kind of stuff that doesn't have to test No.

Now suppose you worry about other things that might have happened to that particular stuff if you hadn't actually done the test you did. Since you actually *did* the test, you might think you've learned something about the stuff which could help you think about the possible results of other tests you might have performed but didn't.

Remarkably, however, if you make the hypothesis that it was a kind of stuff that doesn't have to test No, then there are circumstances under which you can deduce that if you had instead chosen to do a different kind of test, the stuff could have behaved in a way that never happens when that different kind of test is actually done.[3]

---

2 Today (2015) I would say that QBism (see Chapters 30–33) is getting close. But at this point only a few people would agree with me.
3 This is what Chapter 16 is about.

So even though the result was Yes when the stuff was tested, you have to conclude that if it hadn't actually been tested, it might have been the kind of stuff that has to test No. There's an additional assumption you have to make to get to this peculiar position: You have to assume that the character the stuff has can't be altered by a decision made by a friend of yours, who is off in the next county, out of contact with you and the stuff.

Some people conclude from this that the character of your own stuff can indeed be altered by actions taken by your faraway friend. This is called quantum nonlocality. Others, myself among them, conclude that it is treacherous to make judgments about "the character" of stuff, and extremely treacherous to reason from what actually happened to what might have happened but didn't.

It's fascinating to try to write about this.[4] We've known ever since Bohr and Heisenberg, that you can get into trouble if you try to infer the existence of properties independent of the process by which those properties are actually ascertained. But its only through decades of trying to simplify and refine that kind of argumentation that we've come up with examples where you get into trouble by insisting that stuff that has just tested Yes, prior to the test could not have been the kind of stuff that must test No.

The very beautiful and concise mathematical language of quantum mechanics is designed to make it impossible to say things like the naughty things I've just said and, preferably, impossible even to think such thoughts. The mathematical language tells you the possible answers, and their likelihoods, if you ask a particular question about particular stuff. It is incapable of formulating questions about what would have happened under altered conditions based on what actually did happen under actual conditions.

Language, on the other hand, is filled with talk about what might have happened but didn't. If only I hadn't turned left at Cayuga Street rather than Geneva, I would not have been sideswiped by

4 See Chapters 4, 7, and 16.

the car pulling out from in front of the library. Physicists have discovered that there are grave inadequacies built into such forms of expression.

The difficulties only arise when you try to combine what quantum mechanics tells you about the possible results of several different experiments, only one of which can actually be carried out. So one way out is not to worry about what *would* have happened in the experiment you *didn't* do, if you no longer can do it. That's a problem for philosophers, not physicists.

Philosophers of science, however, have by and large chosen to embrace nonlocality as a natural phenomenon, rather than homing in on what bad habits of thought and expression make so implausible an inference so hard to resist. Uncharacteristically for philosophers, they ought to be *more* worried than they are about the nature of language, how it can trap us into formulating questions that have no sensible answers, and whether it is possible to restructure ordinary language in a way that liberates us from those built-in errors that make it so hard to think sensibly about quantum physics. Philosophers ought, in short, to be worried about writing physics.[5]

The challenge of expressing in ordinary language matters whose most natural representation is nonverbal mathematics operates on much less profound levels. At the urging of the National Science Foundation, the Cornell Center for Materials Research has recently become interested in "outreach." Outreach means it's not enough to do research; you should also delight and instruct the general public.

A column called "Ask a Scientist" now appears in the *Ithaca Journal*.[6] It is an example of outreach. I was once asked to answer the following question: "Why is it that when I look at one side of a

---

5 I address what I think philosophers ought to be thinking about in Chapters 30–33.
6 In 2015 it seems no longer to be there.

spoon I see my reflection right-side up, and when I turn the spoon over I see my reflection upside down?"

Reaching out turned into an all-day challenge. It helped not knowing any conventional optics, because I first had to figure out for myself that the spoon behaved as advertised. Then I looked in a spoon to make sure. It did. Then I had to decide what I could take for granted. In this case, how flat mirrors behave. Then I had to find a concise way to express the simulation of a curved mirror by a collection of flat ones. Then I had to find a way to say this in a language of spoons, not mirrors, anticipating and thwarting every imaginable misreading. Here's the result:

To make it easier to picture, think of an enormous spoon, about as big as your head, not counting the handle. You can understand how a curved mirror behaves by thinking of it as built up out of lots of little flat mirrors. So suppose the enormous spoon is a wooden one, made to reflect by gluing a lot of little flat mirrors to both its surfaces, like mosaic tiles on the inside and outside of a dome.

Now imagine holding the spoon vertically some distance from your face, and looking directly into the bowl part of the spoon, with the middle of the bowl at the level of your eyes. As you lower your eyes toward the lower part of the bowl, the little mirrors that you see will tilt upwards, so you see in them the reflection of the upper part of your face. But as you raise your eyes toward the upper part of the bowl, the little mirrors that you see will tilt downwards, so you see in them the reflection of the lower part of your face. In other words you see yourself upside down.

On the other hand, if you turn the spoon so you're looking at the outside of its bowl, then as you lower your eyes the little mirrors that you see tilt downwards and you see a reflection of the lower part of your face, and as you raise your eyes the mirrors that you see tilt upwards and you see a reflection of the upper part. So reflected in that side, you look right-side up.

You may not agree, but that is serious writing. The effort of producing it was similar to what I underwent writing the disquisitions on relativistic and quantum physics in the earlier parts of this essay.

But it's not just quantum mechanics, relativity, and spoons that get me writing physics. I'm convinced that you don't understand the significance of the research you've been struggling with for the past few years, until you begin to write about it. Only then do you realize that it is much more interesting (or, if you're unfortunate and uncommonly honest, much less interesting) than you'd thought. Only then do you really see how your work fits into or, if you are lucky, changes the character of the tradition out of which it grew.

This requires time. You cannot go through such a ruminative process if you feel the urgent need to get your work out ahead of your competitors. Because my written physics has to be slow-aged before it's fit for consumption, I've always sought unexplored backwaters to work in, or obscure corners of otherwise fashionable enterprises. At my back I'd rather not hear the competition hurrying near.

I was talking to a younger colleague about this just a few weeks ago. He was worried that his most interesting work was peripheral to what most people were doing, and therefore failed to arouse their interest. I said that if *he* found it interesting that ought to be enough—that I'd managed to get through an entire career that way. "Ah," he said, "but you've survived by managing to write about it in a way that makes people think it's interesting, even though it isn't." I took that as a compliment. It may be an example of what the organizer of this symposium had in mind when he asked whether we considered our writing to be at odds with our discipline.

A special problem for the writing of physics is created by the predominance of multi-author papers. Research is usually a collaborative process and writing has, inevitably, become highly collaborative. Single-author papers are now rather rare, and papers can be seen whose list of authors constitutes a quarter of the entire text. This is unfortunate. It is hard to discern an authorial voice in such papers. It is now almost impossible to acquire a sense of a physicist's style from a perusal of his or her collected papers, since many people have never in their lives written a paper without co-authors.

This is a tough milieu for one who views the writing as a major component of the research. I once remarked in a *Physics Today* column[7] that "My writing is a process that does not converge; I cannot read a page of my own prose without wanting to improve it. Therefore when I read proofs I ignore the manuscript except to check purely technical points. Proofreading offers one more shot at elusive perfection." An official of the American Physical Society, conjuring up visions of me systematically altering what had passed the scrutiny of peer review, asked whether I fussed that way with my technical articles. He was horrified when I told him that I fussed with them even more.

How can you do that when you collaborate? My solution has been to avoid collaboration. This is easier for a theorist than an experimentalist. With one important exception, my collaborators over the years have been almost exclusively my own graduate students. It's an agonizing process. They, of course, produce the first draft. For some years I would then return a second draft which bore little resemblance to what I had been handed. I recognized the ghastly pedagogy in this procedure, but indulged myself with the thought that they would surely learn much from the contrast between the two versions, and this would show up in the next first draft of the next joint project. And indeed, to some extent this worked. I was brought up short by a student who, I discovered with mixed emotions, took writing as seriously as I did. She was enraged by my second draft. I behaved honorably after that, and we've been good friends ever since. But the paper I wrote

7 Chapter 5.

with her is the only one on which my name appears that feels like
I had nothing to do with it, even though I remember participating
actively in the analysis.

The striking exception to my inability to write collaboratively,
is my eight-year collaboration with Neil Ashcroft on our 800-page
book on solid-state physics. We have very different prose styles.
Yet the book has a clear and distinctive uniform tone, which
can't be identified as belonging to either of us. I think the reason
this worked was that Neil knows solid-state physics much better
than I do. So he would produce the first drafts. Characteristically,
I would not understand them. So I would try to make sense of
what he was saying, and then produce my typical kind of irritating second draft. Neil, however, would now have to correct all my
mistakes in a massively rewritten third draft. I would then have
to root out any new obscurities he had introduced in a fourth
draft. By this kind of tennis-playing, we would go through eight
or nine drafts, and emerge with something that was clear, correct,
and sounded like a human voice. That voice, however, was neither
of ours.

Let me call your attention to one more peculiarity about writing
physics, pertinent to this Cornell lecture series. Humanists read
papers. Physicists give talks.[8] The tradition of talking informally
is so strong that most physicists are shocked when they discover
that people in other disciplines read their talks from a prepared
text. My presentation for the Knight Distinguished Lecture Series
was only the third or fourth time in my life that I've done it. Only
sissies read their talks.

Since the invention of the overhead projector an exception
to this rule has gradually emerged among scientists. It's OK to
read your talk provided you write it on plastic and project it on
a screen so everybody else can see what you're reading. With
this compromise you get neither the precision of the written

8 This difference is discussed at some length in Chapter 42.

language, nor the spontaneity of informal speech.[9] It's an art form that seems to have become particularly popular with university administrators.

In high school I took a test that was supposed to tell me what to do with my life. It was called the Kuder Preference Test. You were asked questions like "Would you rather spend an hour reading to an invalid or taking apart a clock?" You answered the questions by punching holes in an answer sheet with a pin. They told me afterwards that the test showed that I had two great interests: science and writing. So, they said, I should aim to become the editor of a science journal.

Implicit in this recommendation is the distinction made by my English Department colleague, Jonathan Culler, between writing and writing up. Clearly the proprietors of the test knew that scientists produced papers, but evidently they thought that this was writing up—not writing. Writing was done in editorial offices; in laboratories you only wrote up.

But writing physics is different from both writing up physics, and the editorial refinement of written-up physics. While there has to be something there before the writing begins, that something only acquires its character and shape through writing. My transformation of the spoon into a dome with mosaics is clearly not writing up physics. I like to think it is writing physics. The distinction between the two might shed some light on debates in the "science wars" between physicists and social constructivists.[10] The physicists believe that there is a clean distinction between objective truth and mere social convention. They view physics as a process of discovering and writing up objective truths. Social constructivists—at least the ones I find interesting—maintain that objective truth and social convention are so deeply entangled that

---

9 The problem has been made even worse by the replacement of overhead projectors by PowerPoint.
10 See Chapters 17, 18, and 20.

it's impossible to separate the two. For them physics is not writing up. Physics is writing.

Who would have thought, before Einstein's 1905 paper, that simultaneity was a convention—not an objective fact—that clocks were not a useful invention for the recording of objective time, but that time itself was a useful invention for characterizing the correlations between objective clocks? The real issue in the debates of the "science wars" is not whether the physical can be disentangled from the social—the real, from the conventional—but whether their deep entanglement is trivial or profound—a fruitful or a sterile way of looking at the scientific process.

The great Russian physicist L. D. Landau is said to have hated writing. He co-authored an extraordinary series of textbooks in collaboration with E. M. Lifshitz, who did all the writing. From my perspective Lifshitz operated in a co-author's paradise. He was linked to Nature through Landau, who was in deep nonverbal communion with her, but had no investment whatever in the process of articulating that communion.

It is said that even Landau's profound technical papers were actually written by Lifshitz. Some physicists look down on Lifshitz: Landau did the physics, Lifshitz wrote it up. I don't believe that for a minute. If Evgenii Lifshitz really wrote the amazing papers of Landau, he was doing physics of the highest order. Lev Landau was not so much a coauthor, as a natural phenomenon—an important component of the remarkable coherence of the physical world that Lifshitz wrote about so powerfully in the papers of Landau.

So the testers were right about my interests—just wrong about how I ought to exercise them. They ought to have said, "You like science and you like writing, so be a scientist. Write science!"

References

1. N. D. Mermin, *Space and Time in Special Relativity*, McGraw Hill (1968).
2. The new book has been finished: *It's About Time*, Princeton University Press (2005). My 1968 book is still in print. I prefer the new version.
3. *One, Two, Three … Infinity*.

43. Postscript

This essay is based on my Knight Distinguished Lecture in Writing in the Disciplines, given at Cornell in April 1999. It appeared in *Writing and Revising the Disciplines*, Jonathan Monroe (ed.), Cornell University Press (2002), pages 15–28, and was reprinted as a "Guest Editorial" in *American Journal of Physics* **74**, 296–302 (2003). I've cut parts of the lecture drawn from earlier essays in this volume.

INDEX

~

abolition of grants, 75–81
abstractions, 142, 143, 170, 174,
    208–216, 226, 229, 239, 304
accelerator physics, spin-offs, 82
acceptor, 338
action at a distance, *see* quantum
    nonlocality
Afghan hounds, 134, 138, 276
agent, 219, 221, 224, 238
agriculture, 152
Aida, 134, 276
Albright, Madeline J. K., 269
Alexandria, Library at, 65
Ambegaokar, Vinay, 334, 336
ambiguity, 221
American Academy of Arts and
    Sciences, 146
American Physical Society (APS),
    13, 222, 371
    and prizes, 16, 20, 21
analog computer, 306
analog-digital interplay in
    quantum computer, 204
Anderson, Philip W., xii,
    66, 183
apotheosis of informal dress, 271
archeology of physics, 60,
    62, 149
Archimedes, 58
ark rule, 105

arXiv, 72, 89, 172, 181
    actual, 82–89
    anticipated, 67–74
Ashcroft, Neil W, 56, 190, 323,
    347, 372
    not Professor Mozart, 56, 66
assembly language, 314
astonishment, 168
Astounding Science Fiction, 365
atheistic communism, 151
attention, 86
authors, multiple, 34, 70
    voice of, 34
awards, 16–22
Awful German Language, 144

Bachelet, Giovanni, 336
bad taste, 134, 255
baseball, 19, 149, 332, 334
Bayes, Thomas, 226
Bayesianism, 229, 232
    personalist, 219, 220
Beerbohm, Max, 173, 297–300
Beethoven, Professor 51, 76
behaviorism, 311,
belief, 219, 220, 226, 242
    vs. knowledge, 221
Bell, John S., 216, 221, 222, 232–248
    and QBism, 237
    reaction to GHZ, 49

Bell's Theorem, 43, 46, 47, 109,
    175, 245, 304
Berkeley, George, 213
Berry, Michael V., 293
beta decay, 264
bets and probability, 244
beware, 33
bicycle, 232, 245, 293
Big Bang, 55
biology, 303
Birmingham, U. K., 177, 331,
    348–351
blackbody, 190
Bohm, David J., 209,
Bohmian mechanics, 237
Bohr, Aage N., 216, 348
Bohr, Niels H., 23, 25, 28, 45,
    58, 109, 174–179, 192, 213,
    221, 222, 224, 227, 229, 235,
    237, 239, 243, 244, 253, 350,
    356, 367
boiling point of water, 122
Boltzmann, Ludwig E., 159, 284
Boojums, 251–256
bookkeeping device, 170, 212, 308
Born, Max, 181
Born rule, 169, 181, 225
Bose, Satyendra N., 191
Boswell, James, 208
Botticelli, Sandro, 31
Bragg peaks, 155, 156
brainschmaltz, 5
Bridgman, Percy W., 310
British Museum reading room,
    297–300, 308
de Broglie, Louis, 209
Bromley, D. Allan, 75, 81, 82
browsing, 87
Brush, Stephen G., 284

bulletin board, electronic,
    67–74, 82–89
bulwark, 105, 106
Burger, Warren E., 29
Bush, George the First, 21, 81,
    82, 102
    and wave-particle duality,
    100–102
Bush, George the Second, 146,
    300, 317
business, 225
Butterfield, Jeremy, 43

calculate, *see* shut up and calculate
calculator, 256
carbon paper, 96
Carl XVI Gustaf, 133, 273, 275, 277
Carnap, Rudolf, 227, 309
carpet, 148, 152
Carter, Jimmy, 79
Castro, Fidel 82
Caves, Carlton M., 219, 232
CBism, 214, 227–231, 247
CBit, 162, 205, 207
CERN, 64, 265, 266
certain loss, 219
Chandrasekhar, Subrahmanyan,
    154, 283
charming monograph, 30, 32
chauvinism of the present
    moment, 231
Chayes, Jennifer T., 317
    knows how to spell, 4
chemistry, 168, 258, 301, 333, 339
chronosynclastic infundibulum,
    see infundibulum
CHSH (Clauser-Horne-Shimony-
    Holt) inequality, 113, 114
cigarettes, 134, 137, 275, 281

cigars, 82, 86, 87, 148, 151, 151, 251, 257
classical, 213, 239, 243
Clemens, Samuel L., 144
Clifton, Robert K., 43
Clinton, William J., 81, 102
clocks
  atomic, 212
  nature of, 59, 170, 212, 229, 284, 307, 308, 374
clues, 121
CNN, 271, 278
code breaking, 159
cohesive energy, 337
collapse of wave function, 161, 210, 222, 225, 237
Collins, Harry, 117; *see also* *The Golem*
colloquium, 175, 176, 326
color, 261, 262
communism, 148
  atheistic, 151
complementarity, 165
computation, quantum, *see* quantum computation
computational basis, 161
computational complexity, 159
computer consciousness, 311
computer science, 168
computer scientists, 159
concepts, invented, 232
conclusions, 95
condensed-matter physics, 11
  funding problems, 50–56, 103
  origin of the term, 55
conferences, too many, 77
conference proceedings, 77, 86
configuration space, 209
Congress, 151
consciousness, 171, 222, 308

consensus building, 119, 123
conservation laws, 283
consistent histories, 237
constancy of velocity of light, 124
Constitution of the United States, 29
Copenhagen interpretation, 23, 24, 28, 180–186, 192, 232–248
  Peierls disliked term, 246, 355
copy editing, 29–34, 247
Cornell, 10, 72, 123, 322, 331, 333, 341, 362, 375
  Business School Library, 225
  Center for Materials Research, 368
  Department of Chemistry and Chemical Biology, 168
  Department of Physics, 342
  Department of Physics and Metaphysics, 301,
  Department of Science and Technology Studies, 123
  Department of Science Education, 363
  excellent graduate students, 106
  Laboratory of Atomic and Solid State Physics (LASSP), 96, 190
  Nobel Prizes, 137
  Physical Sciences Library, 15, 351,
  Swedish alumni, 132, 271
correlations, 28, 43, 111
cosmology, 57–66
Cosserat continuum, 157
de Coulomb, Charles-Augstin, 283
Cowan, Clyde L., 306
creationism, 147–153
  nature of, 151, 153

creationist assumption, 151
critical point, 344
cryptography, 196, 315
crystallinity, 155, 291, 323
crystallographers vs. physicists,
    155, 160
crystallography in high
    dimensions, 156
cubism, 233
Culler, Jonathan D., 373
curiosity-driven research, 150
curvature, 213

D-brane, 301,
Danish Physical Society, 327
decadence, 312
Democritus, 311
density functional, 332, 338
determinism, 27, 244
Dieudonné, Jean A. E., 154
dimension, 213
dimensionless parameters, 296
diffraction, 98–100
digital-analog interplay in
    quantum computer, 204
Dirac, Paul A. M., 154, 283
discovery, 150
dogs, proud and enormous, 134,
    138, 276
donor, 338
dress, 95
Dukakis, Michael, 21,
Dull, Charles E., 361
Dutch book coherence, 244
dynamite, 273
Dyson, Freeman J., 322
    lifts an eyebrow, 323

economics vs. physics, 132, 271
econophysics, 301

Eddington solar eclipse
    expedition, 124, 306
editorial boards, refusing to
    be on, 14
Edwardian, pronunciation of, 107
Einstein, Albert, 5, 27, 109–116,
    124, 139–146, 160, 187–194,
    196, 212, 213, 215, 222, 227,
    284–286, 291, 296, 302, 304,
    306, 308, 309, 363, 375,
    on the beach, 91, 194
    cites Boltzmann, 160
    locality, 193
    letter to New York Times, 188,
    writing, 191
Einstein-Podolsky-Rosen (EPR)
    experiment, 26, 27, 27, 43–49,
    187, 209, 244
    contrasted with Hardy, 109–116
    immune to charms of, 112
election, presidential see
    presidential election
electromagnetic fields, reality
    of, 211
electron, 257, 263
    diffraction, 98–100
    spinning 157
electronic distribution of papers,
    7, 67–74, 89, 172, 181
electro-weak unification, 62
elegance, 154–160, 283–295
"elegance is for tailors", 160, 284
email, 66, 67, 68, 72, 82, 342
emergency decree, 230
emotion, 34
empiricism, 233
energy bands, 338, 349
engineers vs. physicists, 157
English cooking, 349, 350
English language, 103

Enoch Soames, 173, 297–300, 305
ensemble, 219
entropy, 286
Eötvös experiments, 306
EPR reality criterion, 43–49, 244
equations
    ending a quotation, 35
    glowering and menacing, 37
    number all, 36, 40, 41
    punctuation of, 38
    referring to, 37
    unpunctuated, 39
    writing of, 35–42
eta meson, 262
ether, 215, 226, 303, 304
Euclidean algorithm, 198
Eugene Onegin, 259
event, 211, 228, 229
evolution, 153, 215
experience, 176, 177, 213, 214, 220,
    221, 222, 224, 226, 227–231,
    233, 240
external, 224, 228, 233, 234
eyes glaze over, 203

factoring, 172, 195–200
facts, 161
fairies and witches, 226
Faraday, Michael, 283, 307
Faust, 136, 281
fearful symmetry, 152
feasibility of quantum
    computer, 159
features residing, 111
features responsible, 111
Feigenbaum, Mitchell J., 254
fellowships for graduate
    students, 76
feminine rhymes, 258

Fermi sea, 338
Feshbach, Herman, 28
Feynman, Joan, xiii, 186
Feynman, Michelle, 256
Feynman, Richard P., 28, 180–186,
    252, 303, 304, 347
    diagrams, 326
    dim view of philosophers, 352
Fifth Amendment, 191
de Finetti, Bruno, 226
first person pronouns, 34, 233, 243
Fisher, Michael E., 40, 285, 321–330
    Fisher's rule, 36, 40, 41
    rearranges hotel furniture, 327
    what would Michael do?, 327
Fitzgerald contraction, 216
Fix, Dovetail, 12
Flamm, Dieter, 285
flavor, 263
Fledermaus, 132, 134, 272
FORTRAN, 314
Fourier analysis, 202–204
framework, 237,
frequentism, 219
    problems with, 225
Freud, Sigmund, 235
frog, smiling and jumping little
    green, 136, 280, 281
Fuchs, Christopher A., xii, 219,
    222, 232, 234, 238, 246
fun, 140, 216
fundamental constants, 296
fundamentalism, 153
fusion, 171, 312

gadfly, 323–325, 329
Galilei, Galileo, 215
Gamow, George, 23, 365
Garland, James C., 90, 95

gauge fields, 264
gauge symmetry, broken, 306
geese, 150
geodesic, 213
geometry and physics, 292
German scientists, fired by
    Hitler, 64
GHZ (Greenberger, Horne,
    Zeilinger experiment),
    43–49, 109
GHz (gigahertz), 49
Gingrich, Newton L., 147
Ginsparg, Paul, 72
glace Nobel, 134
Glashow, Sheldon L., 84
glowing point, 228
gluon, 260, 261
God, 151
    His servant, 330
Goethe, Johann von, 31
golden eggs, 150
Goldstein, Sheldon, 110
The Golem, 117
    letters to the editor, 129
Good Samaritan Rule, 37
Google, 8, 174, 175, 179, 182,
    191, 290,
Google Scholar, 49
Grand Hotel, Stockholm,
    131–138, 269–282
grants,
    abolish all, 75–81
    cut, 50–56, 51, 57, 65, 66,
        67, 69, 76
    to tiger 152
gravity, 291
greatest common divisor, 198
greatest contribution to science,
    84, 89, 172

Greenberger, Daniel M., 43;
    see also GHZ
green plague, 6
Griffiths, Robert B., 321
Gross, David J., xii, 317, 317
Gross, Paul R., 122
Grover, Lov K., 314
G, smoking, 53

de Haas-von Alphen effect, 349
habit, bad, 208–216, 229, 368
hackers, 73
Hall effect, fractionally quantized,
    170, 306
Hamlet, 265
Hardy experiment, 109–116,
    366–368
    why not noticed earlier, 109
Hardy, Lucien, 109
harmonic oscillator, 209
Harvard, 176, 184, 190, 326, 337,
    339, 341, 348
    class of 1956, 341
Hawking, Stephen W., 254
heat capacity of solids, 190
Hegel, Georg W. F., 139
Heisenberg model, 322
Heisenberg, Werner, 25, 28,
    31, 210, 237, 237, 241, 253,
    283, 367
helium-3, superfluid, 131–138,
    157, 269–282
heresy, 237
heroic scientific exposition, 349
Higgs boson, 64, 75, 149, 265
    discovery announced, 266
high dimensional
    crystallography, 156
Hilbert space, 210, 211

history, 95, 352
Hofstadter, Douglas R., 259
Hohenberg, Pierre C., 322, 332, 334
ho-hum, 202
honors, 16–22
Hookes law, 65
Horne, Michael A., 43
    see also GHZ
hotel furniture, rearranging 327
humanists, 93,
humanity, 237, 239, 323
Hume, David, 244
humor, sense of, 141, 146
Humpty Dumpty, 16
hundred-year nap, 167

IBM 704 computer, 314
imagination, 228
inappropriate questions, 25
indirect costs, 11, 76, 75–81
    see also overhead
induction, 244, 245
information, 161, 163
    science, 302
infundibulum, chronosynclastic, 301, 302, 307, 311, 313
inhomogeneous electrons, 338
input register, 162, 201
instruments generate space and time, 142, 145
internal, 226
International Union of Crystallography, 156
Internet, 7, 14, 174, 186
interval, 213
invented concepts, 232, 245
iPhone calculator, 256
irreversible, 221
Ithaca Journal, 368

Ithaca, New York, 108, 266, 323, 328, 333, 345, 352
ITP, 338

Jewish, growing up, 189
job talks, 93
John Paul II, 335
Johnson, George, 196,
Johnson, Lyndon B., 81
Johnson, Samuel, 208, 213
jokes, 17, 59, 65, 141, 144, 181, 183, 326, 335
Jordan, Thomas F., 110
Josephson, Brian D., 252
journals
    editorial boards of, 14
    obsolesence of, 72
    online, 14, 69
    too many 12, 71

Kac, Mark, 344
Kadanoff, Leo P., xii, 343, 344
Kammerlingh Onnes, Heike, 305,
Kant, Immanuel, 11, 139
kaon, 262
King of Sweden, 133, 273, 275, 277
KITP, 339
Kittel, Charles, 349
Kleppner, Daniel, xii
knowledge, 161, 166, 210, 242
    arithmetical, 165
    vs. belief, 221
    of what?, 161, 165, 221, 242
    whose?, 161, 221, 242
Kohn anomalies, 335
Kohn, Walter, 331–340
    of Kohn Hall, 339
    musical CV, 336–339
Krumhansl, James A., 10

Kuder Preference Test, 373
Kuhn, Thomas S., 352

Lagrange, Joseph, 7
Lagrangean, 3–8, 49,
Lagrangia, Giuseppe Ludovico, 7
Lagrangian, 4, 58, 317
La Jolla, 321, 331, 338
Landers, Ann, 11
Landau, Lev D., 194, 237, 239,
    321–330, 346, 374
Landau and Lifshitz, 374
    Peierls on, 239,
    two big mistakes, 239
Lang, Serge, 21–22
Langer, James S., xii, 336
language, 212, 234, 368
    ordinary, 224, 243, 363, 364, 365
laptop computer, 302
Large Hadron Collider (LHC), 64
Latour, Bruno, 139–146, 160
    elected to American
        Academy, 146,
    letter from, 146
laws of physics, 353
lectures, *see* talks
lectures on wave-particle
    duality, 97
Lee, David M., 131–138, 190,
    269–282, 342
Leggett, Anthony J., 342
leptons, 60, 264
Levitov, Leonid S., xii
Levitron™, 292
Levitt, Norman, 122
libraries, 6, 9–15, 70, 71, 74, 85, 86
    disappearance of, 15
life, making harder than
    necessary, 208–216
Lifshitz, Evgenii M., 237, 239, 374

linguistics, 103–108
    silly? 108
linguists, what drives them
    nuts, 108
long division, 198
Lorentz, Hendrik A., 253
Lorentz Professor, 328
Lorentz transformation, 144
Lubkin, Gloria, 11,
luck, good 198

MacArthur Awards, 19
macroscopic, 58, 213, 221,
    229, 240
    convenience, 170, 308
magic, 361, 365
magicians, 344
magnificence, 134
many worlds, 161, 163, 224,
    237, 311
Mark Twain, 144
marriage, 224
masculine rhymes, 258
mass, 189, 361
mathematical coincidence, 256
mathematical physics, 181, 326
mathematical tools, 210, 211, 237
mathematicians vs. physicists, 155
mathematics, writing of, 35–42
Math Is Prose Rule, 38
Matthew effect, 160, 28, 180–186
Maxwell, James C., 259
Maxwell's equations, 283
McCarthy, Joseph R., 192
mean field theory, 306, 321
measurement, 162, 202, 204, 210,
    213, 216, 224, 232–248, 308
    limits to what can be
        learned, 163
    outcome of, 232–248

measurement (*cont.*)
    problem, 220
    QBist view of, 240
    strangeness of, 163
memory, 228
Mendelssohn, Felix, 273
Mermin, Dorothy M., 138, 328, 331, 348
Mermin, Elizabeth R., 21, 141–142, 146
Mermin, Jonathan G., 271
Mermin, N. David
    breath taken away, 256
    declining years, 213
    defects of character, 35
    dreams of, 97, 137
    electronically challenged, 72
    forgetful, 180
    ghastly pedagogy, 371
    growing up Jewish, 189
    lacks gun, 213
    lacks patience, 310
    lacks vision, 314
    lecture-room stupor of, 104
    less subtle than Professor Mozart, 65
    makes Greenberger famous, 49
    makes interesting what isn't, 370
    over his head, 342
    proofreading practices, 29
    prophet, 72
    proud parent, 141, 145
    pundit, 82
    quixotic, 189
    recants, 48
    science, what he learned about, 123
    silly, 108, 204

string theory, view of 84
    woolly thinking, 48
Merton, Robert K., 183
mesons, 258, 262
    ain't much fun, 341
metals physics, 349
metaphysics, 46, 158
Michelson-Morley experiment, 124–130, 353
microscopic, 221
microwave oven, 328
Migdal, Arkady B., 286
Miller, Dayton C., 127, 130
Milman, Howard T., 327
Minkowski, Hermann, 214
misattribution, 28, 174–179, 180–186
missiles, interdisciplinary ballistic, 140
MIVeBs, 75, 80
modular arithmetic, 197
monograph, charming, 30, 32
Moon, journeys to, 312
Morandi, Giorgio, 208
Mother Nature, 30
Mothers for Intermediate Vector Bosons (MIVeBs), 75, 80
Mottelson, Ben R., 216
mouse, 222,
Mozart, William A.
    actual identity of, 56
    against grants, 75–81
    archeology of physics 57–66, 149
    articles by, 91–96, 251–256
    celebrates standard model, 257–266
    cigars, 82, 86, 87, 148, 151, 151, 251, 257

creationism, 147–153
  electronic bulletin
    boards, 82–89
  grant cut 50–56
  leaves physics, 147
  letter from Kazan, 66
  love of particle physics, 65, 265
  more subtle than the author, 65
  not Neil Ashcroft, 66
  supports SSC, 65, 80, 82, 85,
    88, 148
  talks 90
  unmade discoveries of, 54
Mozart, Wolfgang A., 31, 50, 57,
  132, 133, 147, 273, 274
muddle, 166
multiple authors, see authors
muon, 264
myth, scientific, 122
myth, sustaining, 119, 121

National Academy of Sciences
  absurdity of election to,
    17, 21–22
National Science Foundation
  (NSF), 50–56, 348, 368
National Security Agency, 159
natural selection, 152
nature, 30, 215, 222, 225, 306
naughty thoughts, 367
Nazis, 189
neutrino, 263, 264, 306
neutron, 258, 259–260
Newark, 106
Newton, Isaac, 291, 364,
  laws of motion, 351
New York Review of Books, 351
New York Times, 50, 123, 167, 183,
  187, 192, 196, 198, 296

Nietzsche, Friedrich W., 139
Nixon, Richard M., 116
Nobel, Alfred B., 134, 275–277
Nobel,
  banquet, 275–277
  cold, 135
  guest, diary of, 131–138,
    269–282
  Memorial Prize, 138,
  Prize, 339, 342
no cloning theorem, 202
noise, 199
nonlocality, see quantum
  nonlocality
November revolution, 257,
  265, 266
Now, 227–231, 309
  of different people, 230
  psychological width
    of, 230
nuclear engineering, 302
nucleon, 258, 259, 262, 337
number theory, 195–200

Obama, Barack H., 224
object, 221, 225
objective probability, 226
object-subject relations, see
  relations, object-subject
obscenity, shouted, 157
Occam's rule, 36
Ohm's law, 307
Onegin stanzas, 257–266
one-time code pads, 315
Onsager, Lars, 349
opalescence, critical, 190
Ørsted, Hans C., 283, 307
Osheroff, Douglas D., 131–138,
  190, 269–282, 342

outcome, *see* measurement
output register, 162, 201,
outreach, 368
Overhead, Argument from, 11; *see also indirect costs*
overhead projector, 94, 96, 372
Oz, Land of, 258

Pais, Abraham, 187
paradoxes, 215
paragraphs, 29
parity nonconservation, 57
Park, Robert L., 51
particle physics, 57–66, *see also* archeology of physics
Pauli, Wolfgang E., 306, 350
    dim view of philosophers, 352
peer review, *see* referee
Peierls, Genia, 350, 354
Peierls, Rudolf E., 9, 177, 178, 221, 232–248, 331, 346–357
    close to QBism, 243, 245
    dim view of philosophers, 352
    early universe, 238
Peres, Asher, 242,
period finding, 158, 195–200
    difficulty of, 199
    and factoring, 195–200
periodicity in crystallography, 155
Perot, H. Ross, 98–100, 102
Petersen, Aage, 174, 179, 356
PhD, 222
philosophers, 166, 245, 368
philosophy
    Feynman's disdain for, 185
    Pauli's characterization of, 352
    Peierls' skepticism of, 347, 352
    tranquilizing 28, 192

phlogiston, 226
phonon, 338
photon, 264
Physical Review, 4, 6, 9, 12, 14, 29–34, 72, 187, 188
Physical Review Letters (PRL), xi, 3–8, 9, 10, 14, 49, 67, 68, 69, 141
    as ash tray, 86
physicists
    vs. crystallographers, 155
    vs. engineers, 157
    vs. mathematicians, 155
physics
    compared with baseball, 150
    and metaphysics, department of 301,
    nothing to say about Now, 228
    as spectator sport, 19
Physics Today, xi
    allows enthusiasm, 49
    announces prizes, 16
    columnist with, 251
    cuts Nobel Diary, 138
    fails to punctuate equations, 39
    full names, 31
    Mermin-Wilson letter, 10
    nonexistent people, 252
    refuses to publish Mozart's book review, 255
piddling laboratory tests, 241
pillow, quantum 28, 192
pilot wave, 209
pills, 269
Pinch, Trevor J., 117; *see also The Golem*
ping!, 25, 28
pion, 262
Pippard, A. Brian, 348
plagiarism, 86, 28, 180–186

plague, *see* green plague,
 white plague
Planck, Max, 190
Platonism, 165, 211, 236, 321
pledge allegiance to flag, 20–21
plumbers, 284
pneumatic tubing, 313
Poirot, Hercule 140
Popper, Karl, 24, 253
pork, 103–108
Portable Document Format
 (pdf), 73
postdocs, 77
PowerPoint, 96, 372,
pre-existing properties, 109
preprints, 6, 7, 67, 69, 74, 84
presidential election of
 1988, 16, 21
presidential election of 1992, 97
presidential election of 1996,
 133, 274
presidential election of 2004, 180
presidential election of 2012, 225
press, secular, 188, 196
 vs. sacred, 188
Price, Huw, 231
Princeton, 326
prizes, 16–22
probability, 27, 229,
 and certain loss, 219
 frequentist, 219
 not inherent, 220
 $p = 1$, 233, 244
 personalist Bayesian, 219
Professor Mozart, *see* Mozart,
 William A.
promotions 14, 19, 32, 69, 70
pronouns, first person 34, 233, 243
proof reading, 29–34
properties, pre-existing, 109

prose
 mathematical, 35–42
 scientific, 32, 33
proton, 258, 259
Pullum, Geoffrey K., xiii, 108
punctuation, in mathematics, 38,
 115, 116,
Pushkin, Alexander S., 259
puzzles, 26, 57, 63, 97, 140, 159,
 161, 166, 169, 181, 223
Pythagorean theorem, 286–291

qualia, 310
quantum
 computation, 158, 161–166,
  172, 173, 181, 195–200,
  313–315
 differs from classical,
  162, 205
 digital-analog interplay in, 204
 what is actually calculated, 163
 field theory, 169, 210, 211,
  322, 373,
 Fourier transform, 202–204
 gravity, 301
 information theory, 193
 interference, 17
 magic, 48
 measurement problem, *see*
  measurement
 nonlocality, 27, 161, 209, 210,
  225, 245, 304, 367; *see also*
  spooky actions at a distance
 paradoxes, 219
 parallelism, 202
 pillow, quantum, 28, 192
 puzzlement, 161
 state, 25, 161, 202, 207, 208,
  220, 234
 in early universe, 237, 238

quantum (*cont.*)
  not objective, 163, 225, 237;
    see also objective
  theology, 46
  trade-off, 165
  world, 174–179
quantum mechanics, 17, 23–28,
  57–66, 158, 169, 171, 208,
  213, 229, 230, 238, 303,
  308, 365
  boring everyday, 203
  consciousness and, 310
  impossible to invent, 306
  lectures on, 97
  pronouns and confusion, 233
  works all the way down, 62
quark, 60, 171, 224, 259
  bottom, 263
  charmed, 363, 266
  confinement, 65, 261
  down, 260
  fractional charge, 260
  pronunciation of 103–108
  rhymes with *pork*, 107
  strange, 263
  top, 263
  up, 260,
quasar, 224
quasicrystal, 155
QBism, xii, 214, 219–226,
  232–248, 366
  common ground in, 236, 246
  early signs of in the author,
    208–216
  why shocking, 246
Qbit, 162, 201, 201, 205, 207
qubit, *see* Qbit
questions for 22nd century,
  167–173, 296–317
questions, inappropriate, 25

reading talks, 93, 94, 346, 372
Reagan, Ronald W., 64, 101, 253
realism, naive, 119
reality, 161, 166, 177, 208–216,
  226, 229
Redhead, Michael, 43
reductionism, 63, 170
referee, 36, 37, 40, 68, 70, 71, 73,
  80, 83, 87, 95, 140
Reference Frame, 11, 56, 104,
  156, 223
  dark ages before, 183
reification, 208–216, 234
Reines, Frederick, 306
relations, 177, 178, 222, 309, 310
  in classical physics, 309
  object-subject, 225, 230, 235
relativity, 9, 189, 291, 306,
  361–365, *see also* clocks, nature
  of
  golemization of, 124–130
  Latour on, 139–146
*Relativity: The Special and General
  Theory*, 139–146
  cites Boltzmann, 160
religion, 28, 151, 192, 235
Rembrandt, 271
research, strategic, 147, 150, 152
rho meson, 262
rhymes
  feminine, 258
  masculine, 258
ribbit, 137, 281
Richardson, Robert C., 131–138,
  190, 269–282, 342
Rohrlich, Fritz, 28
RSA (Rivest, Shamir, Edleman)
  encryption, 196, 315
rule interaction, 105, 108
rumpus, wild 134

salaries of scientists, 78, 354
Santa Lucia, 135, 136, 279, 280
Schack, Rüdiger, xii, 219, 232, 238, 246
Schrödinger, Erwin 27, 162, 192, 208, 211, 222, 225, 235, 283, 311
Schwinger, Julian S., 337
Science, 140
science, 234
    content of, 244
    critics, 139
    end of, 146
    support of, 75–81
    wars, 117, 122, 139, 373, 374
Science Talent Search, 189
Scientific American, 31
scientific
    knowledge, 352
    method, 119–120
    thinking, 226
scientists
    role played in sciencee, 130
    what was learned this week about, 123
second per second, 229
self nomination, 20
semi-EPR, 111
Sensational!, 113
sexism, 30
Shakespeare, William, 60, 83
shifty split, 221, 240
Shor algorithm, 195–200
    misrepresentations of, 196
Shor, Peter W., 196
shown, it can be 198, 206
shut up and calculate, 24, 28, 180–186
Simon, Mark G., 266
simultaneity, 212, 227, 364, 374

single user, 234
singular events, 69
slavery, 192
sleight of hand, intellectual, 212
Smoluchowski, Marian, 190
snail mail, 17, 21, 254
sneering, 144
Snow, Charles P., 354
social construction of scientific truth, 121, 373
Social Studies of Science, 141
sociological level, 17
sociology, 94, 183, 326, 352
Socrates, 324
Sokal, Alan D., 123, 139
solid-state physics, 288, 348
solipsism, 36, 233
Sommerfeld, Arnold J. W., 225, 230, 235
Soviet Union, 148
space and time
    absolute, 226
    are abstractions, 143, 170, 211, 212, 229, 307, 308
    foam at Planck scale, 213, 307
spacetime diagrams, 214, 229
spinning electron, 157
spooky actions at a distance, 45;
    see also quantum nonlocality
spoon, 369
Sputnik, 331
SSC *see* Superconducting Super Collider
Stalin, Joseph V., 152
standard model, 103, 106
    celebration in verse, 257–266
Stapp, Henry P., 110
state, *see* quantum state
statistician, 226
strategic plan, 152

strategic research, 147, 150, 152
string theory, 84, 167, 171,
    172, 312
  greatest contribution to science,
    84, 89, 172
  utopian, 84
stuff left behind, 109
subject, 221, 222, 225, 228, 235
summer salaries, 78
Superconducting Supercollider
    (SSC), 63, 65, 66, 85, 88, 101,
    102, 148, 266
  cancelled, 64, 81
  international support for, 148
  rally in support of, 75
superconductivity, 85, 170, 171,
    305, 312
  impossible to predict, 306
superstitious beliefs, 226
supersymmetry, 301,
surface tension, 191
symmetries, 284
symmetry, fearful, 152
Szymborska, Wislawa, 131, 133,
    134, 270, 274, 277,

talks, 77, 90–96
  boring and confusing, 92
  dress for, 95
  job talks, 93
  too easy, 91
  written, 93,
tapestry, 121, 122, 128
tauon, 264
teaching, 352
tea leaves, 191
tear gas, 75, 79
technology, *gedanken*, 161,
    166, 172

technology transfer, 78
telephone, 73, 82, 342
  paying for, 80
  pocket supercomputer 74,
  superior to email?, 72
temptation, 109–116
tigers, 147–153, 152
tilings, 288
time
  dilation, 230
  hidden presence of, 364
  nature of 59, 170, 215, 304, 374
time-reversal symmetry
  broken, 57
top quirk, 105, 106
transparencies, 94, 96, 251,
    346, 372
tranquilizing philosophy, 28, 192
Trigg, George L., 255
triumphalism, 120
tunnneling, 365
Twain, Mark, 144
twin paradox, 230
two-slit diffraction, 98–100, 161
Tygers, 153
type of particle 111
typewriter, 96, 326

Uhlenbeck, George E., 285
Ulfbeck, Ole C., 216
unexperienced experiences, 242
unitary transformation, 162
universe, age of, 59
universities, support of 75–81
unperformed experiments, 242
unspeakable, 221
update, 222, 224, 226
Updike, John H., 259
user of quantum mechanics, 238

vagueness, 221
valence bonds, 338
Valhalla, 342
vanity, 88
variational methods, 337
variety of particle, 111
Vatican, 336
Vickrey, William S., 135, 274, 277
vita enhancement, 12
volley ball, 331

Waffletron, 101
Wagner, Herbert, 322
War and Peace, 95
war rule, 102,
    powerful, 106
warx rule, 107
wave function, *see* quantum state
wave packets spread, 209
wave-particle duality, 97
waves and quantum
    factoring, 195
Waxahachie, 64, 66, 88, 102
weak interactions, 62, 264
Weinberg, Steven, 139, 207
Weisskopf, Victor F., 28, 108,
    175–178, 356
what would have happened, 367
white plague, 6, 7
W-meson, 264
who's that guy up there, 335, 336
Widom, Benjamin, 322, 343, 344

Wigner's friend, 220, 242
Wilczek, Frank A, xii
Wilson, Kenneth G., 10, 72, 252,
    325, 341,
    cutest Prof, 341
    rolling ball uphill, 343
Wilson, Robert R., 266
World Wide Web (www), 73, 284
words, bunches of, 61
writing, 7, 14, 29, 191, 355,
    361–375
    Ashcroft and Mermin, 372
    Bohr's, 174
    Kohn's, 335
    madness to go on, 37
    mathematics, 35–42
    multi-author papers, 370–372
    Peierls', 346–357
    physics, 35
    serious, 370
    vs. writing up, 373
Wunsch fulfillment, 354

X-ray photographs, 155, 156

younger than your children, 362

Zeilinger, Anton, 43,
    see also GHZ
Ziman, John M., 348
Z-meson, 264
Zurek, Wojciech H., 324, 325